THE CRAFT OF JEWELLERY MAKING

A COLLECTION OF HISTORICAL
ARTICLES ON TOOLS, GEMSTONE
CUTTING, MOUNTING AND
OTHER ASPECTS OF JEWELLERY
MAKING

BY

VARIOUS AUTHORS

Copyright © 2013 Read Books Ltd.
This book is copyright and may not be
reproduced or copied in any way without
the express permission of the publisher in writing

British Library Cataloguing-in-Publication Data
A catalogue record for this book is available from the
British Library

CONTENTS

INTRODUCTION TO GEMMOLOGY 1

THE JEWELLER'S ART 5

CRAFTSMANSHIP. 23

GUILDS, AND THE INFLUENCE THEY EXERCISED 39

THE ENGRAVER 52

SPECIAL LAPIDARY TECHNIC 60

USEFUL LAPIDARY NOTES 145

ORIENTATION 185

METALS USED FOR GEM MOUNTINGS 249

THE LAPIDARY 370

Introduction to Gemmology

Gemmology is the science dealing with natural and artificial gems and gemstones. It is considered a geoscience and a branch of mineralogy. Although some practice gemmology as a sole profession, often jewellers become academically trained gemmologists, qualified to identify and evaluate gems. Rudimentary education in gemmology for jewellers and gemmologists began in the nineteenth century, but the first qualifications were instigated after the 'National Association of Goldsmiths of Great Britain' (NAG), set up a Gemmological Committee for this purpose in 1908. This committee matured into the 'Gemmological Association of Great Britain' (also known as Gem-A), now an educational charity and accredited awarding body, with its courses taught worldwide. The first US graduate of Gem-A's Diploma Course, in 1929, was Robert Shipley who later established both the 'Gemmological Institute of America' and the 'American Gem Society'. There are now several professional schools and associations of gemmologists and certification programs around the world.

The first gemmological laboratory serving the jewellery trade was established in London in 1925, prompted by the influx of the newly developed 'cultured pearl' and advances in the synthesis of rubies and sapphires. There are now numerous Gem Labs around the world requiring ever more advanced equipment and experience to identify the new challenges - such as treatments to gems, new synthetics and other new materials. Gemstones are basically categorized based on of their crystal structure, specific gravity, refractive index and other optical properties such as pleochroism. The physical property of 'hardness' is defined by the non-linear 'Mohs Scale' of mineral hardness. Gemmologists study these factors while valuing or appraising cut and polished gemstones. Gemmological microscopic study of the internal structure is used to determine whether a gem is synthetic or natural, by revealing natural fluid inclusions and partially melted exogenous crystals, in order to demonstrate evidence of heat treatment to enhance colour. The spectroscopic analysis of cut gemstones also allows a gemmologist to understand the atomic structure and identify its origin; a major factor in valuing a gemstone. For example, a ruby from Burma will have definite internal and optical activity variance as compared to a Thai ruby.

Gem identification is basically a process of elimination. Gemstones of similar colour undergo non-destructive optical testing until there is only one possible identity. Any single test is indicative, only. For example, the specific gravity of ruby is 4.00, glass is 3.15-4.20, and cubic zirconia is 5.6-5.9. So, one can easily tell the difference between cubic zirconia and the other two; however, there is overlap between ruby and glass. And, as with all naturally occurring materials, no two gems are identical. The geological environment in which they are created influences the overall process, so that although the basics can be identified, the presence of chemical 'impurities' and substitutions along with structural imperfections vary - thus creating 'individuals.' Having said this, the three main methods of testing gems are highly successful in proper identification. These are:

Identification by refractive index - This test determines the gems identity by measuring the refraction of light in the gem. Every material has a critical angle, at which point light is reflected back internally. This can be measured and thus used to determine the gem's identity. Typically, this is measured using a refractometer, although it is possible to measure it using a microscope.

Identification by specific gravity – This method, also known as 'relative density', varies depending upon the chemical composition and crystal structure type. Heavy liquids with a known specific gravity are used to test loose gemstones. Specific gravity is measured by comparing the weight of the gem in air with the weight of the gem suspended in water.

Identification by spectroscopy – This technique uses a similar principle to how a prism works, to separate white light into its component colours. A gemmological spectroscope is utilised to analyse the selective absorption of light in the gem material. Essentially, when light passes from one medium to another, it bends. Blue light bends more than red light. Depending on the gem material, it will adjust how much this light bends. Colouring agents or chromophores show bands in the spectroscope and indicate which element is responsible for the gem's colour.

THE JEWELLER'S ART

A JEWELLER—MATERIALS OF WHICH
JEWELLERY IS COMPOSED—NATIVE ART—
CULTIVATED TASTES—DISTINCTIVE PERIODS OF
PRODUCTION—SEATS OF THE INDUSTRY.

IN the previous chapter the review of the precious metals and the methods adopted to secure them suggest the common names of the workers in these metals, and also give the familiar phrase, "gold and silver plate." These workers, however, operated larger things and many objects of utility, as well as fashioning the more important works of art which have given such prominence to the goldsmiths and silversmiths of all ages. Here we must consider their smaller and yet equally as artistic and costly works which come under the head of "jewellery." At first sight it seems difficult to differentiate between a goldsmith or silversmith and a jeweller. This difference is, however, easily distinguishable when the two essential elements of the crafts are considered. The goldsmith works in gold and shapes and fashions it, as the silversmith hammers and chases silver; but when either of these workers in precious metals take up jewels and design or execute a frame or setting for the stones they have selected, or use precious stones for the embellishment of the silver or gold work they have in hand, then they become jewellers. The art of the jeweller has been apparent at all times, and under almost every condition of civilisation the art has been practised. In this chapter it is intended to show the product and skill of the artist, rather than the craftsman as a worker; he must be considered from a different standpoint.

A JEWELLER.

A jeweller then is a man who works in precious stones and upon other objects which he embellishes with a beautiful setting, and thus securely combines the pleasing effects of the sparkling gems and the pure gold or silver, and in earlier days bronze, for we must never forget that the jewellery of prehistoric peoples and of the more cultivated Greeks and Romans was chiefly of bronze, a compound metal in which tin from Britain was employed.

The jeweller must be an artist and a designer before he can excel in his work; and the work performed in the past often shows the characteristics of the jeweller, who stamps upon his handiwork his mark—the mark of his skill and of the peculiar treatment he was wont to impart to the work he had undertaken. The designer and the artist are inseparable. Either the one makes the pattern for the other, or the artist in metal work first makes the design and then executes it. It is only the very crudest design that can be evolved as the work proceeds.

In the making of jewellery in which so much costly material is involved the artist has to take every care of the stones with which he is entrusted, and he has to economise the amount of gold or other metal required in their employment. Jewellery has to be of sufficient strength to withstand the wear and tear of many years, and the artist who makes it must know all about the relative strength and wear of the materials he employs and also of the strain likely to be put upon the different objects when worn.

The jeweller prefers gold, as in most cases it renders him the best results, and is more effective as a setting for jewels rare and beautiful; its ductability too, is in its favour.

Precious stones are of different colours, and used in many settings, some of which are more effective in silver than in gold; and the different effects should be understood by the jeweller who works to produce the best possible results, rather than merely to obtain payment for his work. The credit of doing good work was one of the delights of the craftsman of olden time, in the days before commercial jewellery was made by "the dozen" on stereotyped lines and by machines which duplicated the objects with provoking exactitude. In olden time the work of the goldsmith and the silversmith were more closely allied then they are now, and the jeweller worked in both metals, often the same craftsman operating both metals with equal ease—it is only in modern days that the workman has been confined to limitations, and his range of work limited to set grooves, with the result that evenness and regularity and the following of approved styles have spoiled the natural art of the craftsman of former days, who was then rather an artist than a workman.

In this volume of the "Home Connoisseur Series," ancient domestic plate—silver and gold, and silver overlayed with gold—is not dealt with, only the work of the goldsmith and the silversmith as applied to jewellery and trinkets.

THE MATERIALS OF WHICH JEWELLERY IS MADE.

It has already been shown that the jeweller is a departure from the simple craftsman who worked in gold or silver without the additional stones or other materials of which jewellery is composed. A collection of old jewellery, however, reveals many materials employed in the manufacture of the ornamental and decorative jewellery of past days. Gold it is true has at times been almost exclusively used without stones or gems, as in the case of the Greek jewellery which consisted chiefly of beaten gold. In the Greek goldsmiths work, however, there was a distinct type of decoration, in that the beaten form was covered with much decorative work made of fine wire wrought into delicate patterns.

Filigree work has been wrought in many countries, and especially in India, by native workers. It is of course jewellery without jewels, just as is some of the beautiful lace-like filigree work in silver which is so much admired; the skill of the worker is fully demonstrated in metal without gems. In earlier days bronze was used. Copper has been the foundation of much jewellery that has been plated over. The alloy of cheap gold, generally used, is some form of brass of which, of course, copper is the base.

Sometimes rarer metals, some of which like platinum are more costly to procure, are used either in conjunction with gold and silver or alone. Some of the early rings were massive and consisted of copper only. The materials from which the frames of jewels are made are sometimes composite like the backs of brooches in which are cameos, stones, porcelain gems, mosaics and enamels. This last named material has been very popular during the past few years, although it is but a revival of a much earlier art.

The collector is often at a loss to make quite sure about the substances of which the objects he admires or possesses are made, or the gems set therein, it is therefore well to be familiar with the materials. This is not always easy when the gems are of a somewhat unusual colour or shape, a little practice, however, trains the eye to recognise the stones more commonly met with in old jewellery. There are the diamond, ruby, emerald, garnet, and so on. Then pearls which cannot be mistaken for any other gem (except imitations of the genuine which have been brought to such perfection). There are tests which can be applied to metals to ascertain their purity, for pure gold, because of its soft nature, is seldom employed without alloy to make it firm and lasting. The jewellery of the savage, of the prehistoric Briton, and of the more cultivated Saxon and other early peoples who wore jewellery before they received their tuition from the Eastern races, was made of ductile metals only, and much of the gold used was pure, hence its softness. It answered the purpose of these early artists because it could be hammered into shape, first by stones and afterwards by hammers of bronze.

The plates of gold and pieces of metal used by Anglo-Saxon jewellers typified the simple combination of two well understood materials used in conjunction, the one forming a setting for the other, and by contrast enhancing the effect of the article, which if it had been made from one material alone would have been without style or appearance. Throughout the ages the materials employed have been the same with but slight variations; the introduction of some new material as a setting, or with a view to improving the effect of the simpler combinations. The chief difference between ancient and modern art lies in the craftsmanship, and in the tools the workers were able to bring to bear upon the raw materials, together with the addition of science in the finish of the product.

NATIVE ART.

When we speak of native art it is understood to mean the simple natural productions which man has at all times been able to accomplish without any trained instruction, and without that knowledge of production which comes from serving an apprenticeship to one who has already learned the mysteries of the craft he practised from some one who has in his turn added to the earlier forms of art. The native art of men untutored in either art or craftsmanship is intuitive and inborn, it is man using the powers within him for the first time, struggling still on the first rungs of the ladder of art and knowledge.

The natives of many early races worked in the materials which came to hand and accomplished much without the aid of tools. We can form some idea of the work of a man untaught when the amateur tries for the first time to handle simple tools and aims at copying some old piece of jewellery. He finds his chief success in copying the handiwork of the prehistoric savage.

Englishmen have from time to time had opportunities of seeing native workers in precious metals accomplish much from simple tools and a few materials, but these have generally been the picked workers of the tribe and therefore their work is above the average of the race to which they belong. Those who have visited the great industrial exhibitions which have been held in London during recent years have lingered long before the stands of native jewellers from India, Ceylon and Eastern countries. They have seen these people cunningly fashion with very primitive tools gold and silver jewellery and inset precious stones just in the same way the ancients did.

African natives have shown us how they can twist and work metal wires into bangles and rings and how they are able to use their fingers in this delicate work. Travellers from some of our Colonies, and from South America, tell of their visits to the shops of jewellers where they have seen them working just the same as their ancestors did hundreds and even thousand of years ago, fashioning much the same works of art. Visitors to the East tell too of the way in which they have been defrauded, for now and then they have come across makers of so-called antiques; forgeries of simple objects which can be copied so easily are being made to-day to satisfy the craving for relics and for mementoes of those ancient peoples who lived in Egypt and other places of interest, full at one time, if not now, of relics of the past—links with former generations.

Just as those who live where once ancient civilisations dwelt the natives of many islands and out of the way places work to reproduce copies of the past—native art following with a curious exactness the same arts practised long ago. As an instance the natives of Manilla are great workers in gold and silver, their women making most of the jewellery and trinkets they sell. They are adepts at making necklaces of coral; some of the coral rosaries and strings of beads being enriched by pendants of pearls and filigree gold. The native gold they use is a deep yellow colour, and this they carve and often set with jewels. A clever piece of work is the fashioning of ropes of gold made in imitation of manilla rope or cord. These and other natives are adepts at colour work, and have some "trade secrets" in the preparation of enamels.

In copying native works the amateur and the copyist of antiques is at an advantage in that he has beautifully made tools—steel hammers, plyers, drills, and the like. Most of these tools, however, have their prototypes in the simpler tools of the ancients, and from them they reached the same results, but by much more laborious methods. Instead of using gauges and measuring rules the old workers used to work by "rule of thumb," and depended upon their sight and touch to duplicate their objects and to make some uniformity in their work. These facts are worth noting, for without their recognition it would be sometimes difficult to distinguish between genuine antiques and those forgeries with which the market is flooded.

CULTURED TASTES.

When considering the art jewellery of different peoples it is well to note that when native craftsmen learned from those better skilled in the use of tools than they were, they were able to produce greater fineness of detail in their work than hitherto, and as the tastes of their patrons became more cultured and refined there was a change in style, and a departure from the barbaric effects formerly prevailing. The degrees of culture which different nations have reached cannot be measured by time nor by their association with other peoples, yet whatever form their culture took it is reflected in the art of the period. The art of ancient Greece has never been excelled, for at that time the cutting of intaglios and cameos reached a high pitch. To examine some of those beautiful gems which are to be seen in the National Galleries, and in lesser numbers in private collections, reveals skill truly marvellous. To have been able to produce such minute replicas of statuary and larger works of art shows an appreciation of detail in a great degree.

The Eastern peoples who loved coloured textiles and rare jewels coated with bright coloured enamels had a taste quite different from the Greeks. Then again the Celtic jewellery was artistic in a way, but it did not show much culture or taste in ornament. Later the Saxons had much beautiful gold, and the ornament was delicate and chased in much more refined taste. There is culture in the Indian ornament, but different again. Look at the Indian wood carvings and tracery and then at the gold jewellery, and in the perfection of the latter there is an evident attempt to follow the art which the wearers of jewellery would appreciate and understand. There have been times when the cultured "upper ten" in England have been very loud in their tastes, and a superabundance of jewellery has been popular, it is, however, at the periods in this country's art when culture was most marked that the best jewellery was made. These periods must be traced separately, but as showing the jeweller's art as represented by his works accomplished during given periods, these special times may be given here.

DISTINCTIVE PERIODS.

The changes in a nation's taste are generally brought about by some dynastic changes or by great upheavals, even wars of great magnitude and lasting a long time have a strong influence on fashion and style and in the quality of art work as well as on its design. To explain the way in which these changes are brought about it will be sufficient to refer to the craftsmanship of this country. The crude art of the early Briton was changed by the long occupation of Britain by the Romans. Roman art became the taste, and its style dominated the earlier art of the natives who were taught a different way of working metals.

We admire the Celtic jewellery which is so distinctly designed after the art which is seen in the runes and carvings on the old crosses and ornaments of that period. The art then practised gradually developed into the Mediæval. It was then that jewellery followed the designs and colourings of the furnishings of the ecclesiastical buildings which culminated in the Gothic. Then the goldsmith wrought wonderful jewelled ornaments for abbey and cathedral, and the domestic plate and jewellery followed the same lines.

Tudor influence has already been referred to. When James VI. of Scotland ascended the English throne it is not surprising that the thistle and all that appertained to Scotch ornament was introduced into the design and decoration of jewellery worn at the Court. The style developed during the Stuarts. Then came the break when Puritanical ideas prevailed. Jewels and plate—those not melted down in the Royalist cause—were put away to be remade or altered into the florid style of the Restoration art.

Not only did fashion in jewellery alter according to prevailing styles in architecture and art, but the taste for wearing jewels was encouraged or discouraged by leading ecclesiastics and crowned heads according to their fancy. There was a great revival during the reign of Henry VIII. and the two Queens, his daughters; and at the Court of Elizabeth the wearing of jewels was carried to excess, the costume of the Virgin Queen was a blaze of diamonds and other precious jewels. (*See* Chapter XXXIV., "*Royal and Ecclesiastical Jewels.*")

SEATS OF THE INDUSTRY.

The manufacture of native jewellery was of course common in most countries even at an early date. Peasant jewellery, as it is often called, was to be met with everywhere before any special centres of the industry had been founded. Yet even in olden time certain places became famous for the making of jewellery, their fame spreading as intercourse between countries extended. The Egyptians were clever in their day, and their hammered work became notorious. The jewellery of ancient Troy, and later of Italian cities was distinctive. The wonderful examples of Etruscan jewellery which have been discovered show that there was an art developed there to a great extent. Russian and Spanish jewellery at a much later date were well defined, and showed an established industry in those countries. In more modern days Vienna and Paris have been leading European markets from which noted jewellery has been obtained.

As already indicated much English jewellery was, and is made in London, chiefly in Clerkenwell, a district where foreign workmen settled after the Revocation of the Edict of Nantes. It is said that at one time nearly two thousand persons found employment in this neighbourhood in making jewellery, and in more recent times processes in which the use of machinery has been employed have been in vogue, intervening to prevent that individuality of workmanship observable in the older work.

The great centre of the jewellery trade now is Birmingham, where not only cheap articles but much fine work is made.

The city of Birmingham has been so closely associated with the manufacture of many of the things which are classed as trinkets, as well as jewellery itself—both cheap and of better quality—that it seems fitting in a work of this kind that some direct reference should be made to the productions of that town which has so often been dubbed "Brummagem," in slang parlance. Birmingham gradually became one of the great workshops of this country, and at a more remote period, when only a village, it was held to be the "toy-shop" of the world. In strictly trade terms the manufacture of "steel toys" *was* carried on very extensively. But to the dealer in hardware the "steel toys" meant something very different from nursery toys. It was the trade expression meaning steel and iron oddments, mostly highly wrought and polished, which were added to buckles and chains and trinkets, besides distinctive jewellery. There were many things turned out of the small cottage workshops of the heaths and the villages round about Birmingham doing great credit to the village craftsmen; among these were chased oddments which were attached to the chatelaines of the wives of the eighteenth century and the early years of the nineteenth. Things which are regarded with delight by their granddaughters and great granddaughters in the present day. These things are among the treasures of the home connoisseur.

It is difficult now to realise the quaint old town with its black and white timber built houses, and the picturesque scenes which were enacted at holiday times and on feast days. The men and women of that early manufacturing town were rough but good hearted, and their sports too, were on the true old English type. Such pleasures as bull-baiting and cock-fighting are recalled by place names like the "Bull Ring," now a thoroughfare in that busy city.

Birmingham was chiefly noted for so many small articles that it is wonderful how it prospered; even when much of its business was confined to buckles, buttons and the like, the trade grew, and the steel "toys" for which Birmingham had become famous were in great request. Buckles were made in every possible quality, and following the fashion of the day they were large; they were of steel and silver, and some were of plated metals like shining gold, although not actually made of the precious metal.

The buckle as an article of dress—or dress ornament—was in use as early as the fifteenth century. Then for a time buckles fell into disuse, to be revived in the eighteenth century, when they were worn on shoes; the size was increased until the fashion became extravagant, and some very ridiculous buckles adorned the shoes. Again the metals of which these later buckles were made, varied, for in the closing years of the eighteenth century several new alloys were introduced. One of these was known as "Tutania," called after its inventor Tutin. Buckles are used to-day, in moderation, and not a few ladies are wearing old buckles of almost priceless value, wearing again choice antiques.

It is probable that no one article has been made in such

countless numbers, or in such great variety of size and form, as the modest button. When we think of the different colours, and the varied materials of which buttons are still made, and then look back upon the altering fashions which brought a demand for some new class of button it is not to be wondered at that an assembly of buttons of all kinds would be a very extensive collection, if not a particular fascinating or "brainy" pursuit.

Birmingham was responsible for many of the early buttons mostly of metal, used in such quantities. It must be remembered that there was once a time when every person of note employed servants and flunkeys, dressing them in liveries adorned with shining buttons. Some of the gilt varieties were very ornamental and not a few were decorated with the arms or crests of their owners. These too, came from the great "toy shop."

Perhaps one of the best known factories in Birmingham in the eighteenth century was that of Matthew Boulton of Soho Works; it was there that many important objects were made, and there too, that the "Mint" was set up, producing so many medals and souvenirs. Referring to these old works, in *Old and New Birmingham*, it is said, "Matthew Boulton established himself on Snow Hill as a manufacturer of 'toys,' buckles, clasps, chains, and other trinkets, which exhibited good workmanship joined to artistic design, worked out by the best men he could procure. It has been said of him that he could buy any man's brains, and in this lay the great secret of his success." Here then we have the term "toys" explained, and also learn that among other notable men of Birmingham Matthew Boulton did not despise the making of small things, for in his great workshops he turned out "steel toys" of every known variety.

Birmingham and the Black Country continue to turn out trinkets in countless numbers, but the trade of that great manufacturing area is far-reaching, and the manufactures of the district include immense works of iron and steel for which this country has become famous. Very different indeed is that large manufacturing district from what it was when men worked exclusively in their own little workshops. In those days families became specialists, and the peculiar skill they attained was handed on to succeeding generations. Some would be able to inlay, others to engrave, and some to cleverly fashion those fanciful ornaments which were so evident in the large brooches then in vogue, and which are now worn once again—as souvenirs of the past.

In other parts of England there have been localities where noted objects have been made. Mr. Wallis, in *British Manufacturing Industries*, treating upon "Jewellery," mentions some special things made at Derby. These, he says, consisted of "neatly designed pins, studs, brooches, and rings of a peculiar style of setting, still known among the seniors of the jewellery trade as the 'Derby style'." That was in the seventies, and the style is now almost forgotten. Some of the old traders' catalogues mention these goods, and several trade cards and bill heads of the eighteenth century mention "Derby" jewellery.

As an instance of the light thrown upon the sales of that day a large trade card or bill of George Dean of the "Corner of the Monument Yard, on Fish Street Hill, in London," records some of the trinkets which came from Birmingham and other centres of production early in the nineteenth century. Among other things mentioned are "Gold and Silver Jewellery of all sorts, Buckles, Buttons, Combs, Key Swivels, Etwees, Watch Keys and Seals, and various other articles." John Moore of "Air Street in Piccadilly," on his card dated 1789, announced that he made "Silver and Steel Cockspurs, and Buckles, in the neatest manner." On a bill, dated 1790, George Smith of Huggin Lane announced that he was a "Buckle, Spoon, and Tea-Tongue Maker"—and thus examples could be multiplied indefinitely.

Of the minor local industries mention may be made of the famous jet jewellery of Whitby, sold as souvenirs of visits to those parts of England then less accessible than now. Jet jewellery was also much worn at one time with mourning. Most of the jet ornaments and jewellery of Whitby were, however, made in Birmingham and only ornamented at the place where the material was found in abundance.

CRAFTSMANSHIP.

EARLY ASPIRATIONS—SOME TECHNICALITIES—
COMMON PRACTICE—AMATEUR REPAIRS—
SIMPLE TOOLS—THE RESULT.

THE foreign workers who settled in London, their descendants who became Britishers, the sturdy men of Birmingham, and the best artists who have been reckoned among the cleverest craftsmen of their day, have all attained proficiency after years of hard work. There seems to have been implanted in man a desire to succeed, and in whatever sphere of labour he finds himself he tries to do his best; if not there must be something wrong about the man himself, for there is a natural competitiveness about the human race which prevents a perpetual standing still. There is something within man which compels him to move forward; and in the race some go ahead of their fellows, others lag behind, the nation as a whole, however, goes on towards its destiny: if on the up grade to a glorious future, if on the down grade to disastrous failure. We have seen this continually in nations, and it has been observed in trade and commerce. Art enters the world of production in almost all cases, and this has been very noticeable in the art of the craftsmanship which has produced so much and so varied jewellery, representing every race from the dawn of civilisation onwards.

EARLY ASPIRATIONS.

The early craftsmen gradually acquired proficiency in the arts they practised after much painstaking labour, and, no doubt, many failures. They groped their way towards that perfection which the true artist deems his goal, but which he rarely if ever reaches. In the days when primitive craftsmen were making and fashioning bronze and pure unalloyed metals they worked without any past on which to build, simply trying to shape the article of jewellery or other object they were making so as to combine convenient wear and that degree of beauty to which their aspirations soared, or endeavouring to reach the goal towards which they moved either unconsciously or goaded on by those for whom the jewels were intended. Perhaps even in those early days they were sometimes urged on by competition, which in its nobler form has always been helpful in the betterment of craftsmanship.

Collectors and wearers of antique jewellery rarely concern themselves with the way in which it was made, yet the methods adopted by different races and by men influenced by various surroundings have had much to do with the results achieved and the lasting effect of their work. Some native work, although very crude, has a simple dignity about it which appeals to the connoisseur of art, for the true admirer of art looks rather to the motive and the aim and the aspirations which have actuated the worker than the actual result, when compared with art produced under more favourable circumstances.

The success which attended the craftsmanship of the primitive peoples as evidenced by the relics which come to us from prehistoric tombs, from savage races, and from the untutored natives of the islands of the seas shows that art inspires the worker, and that independent of competition and the tendency to copy the true artist aims at originality, and that in whatever grade he is found success is assured, for native art carries with it that which can never belong to machine-made jewellery however attractive it is.

The same ambition which fired the first workers in metals who attempted the making of jewellery makes the best artists of to-day enthusiastic, and hence it is that the patron of art who has the means and is willing to fully recompense the artist can secure original beauty to-day. Such works of art—the triumphs of the goldsmiths and workers in precious metals and gems are to us what the simple objects of antique jewellery, which are to-day treasured rather because of their antiquity than their beauty, were to the first wearers of those ancient gems.

The apprentice who in Mediæval days had got over the

drudgery of his apprenticeship and was allowed by the master craftsman to work in the precious metals was watched very closely, for before he could become a master hand he would have to fulfil the requirements of the Guild by whom the work of his craft was controlled. He, too, would have early aspirations. He would see in his master's workshop many very beautiful things, the result of experience and practice, for both these attributes are necessary to success. Experience teaches the way to do work, and it gives the confidence which is essential when working in valuable materials.

Practice enables the workman to accomplish his mission, and the two in combination make it easy to carry through any great work which is entrusted to the craftsman. It must be remembered that in the Middle Ages there were comparatively few workers in any one art craft. The ecclesiastical support given to art enabled many of the best jewellers, or goldsmiths as they were more generally called, to undertake large and costly works of art in which many rare gems were cut and polished and set in appropriate framework, and what is perhaps of more importance adapted to some real use either in civic or church purposes. The wealthy nobles have in all ages, and in all countries, been the patrons of art—art as they understood it to be—and to the moneyed class is due the success achieved by the men with aspirations and abilities, but with little capital of their own to become possessed with the materials on which to work.

SOME TECHNICALITIES.

There was a beginning to all arts, and although it may not always be very clear where the commencement of any given period can be placed, it is generally found that while native jewellery is always crude, and at times barbaric, it is seldom that we are able to get down to the rock bottom. The jewellery found in the graves of the earliest period of the Bronze Age shows signs of some degree of proficiency, and must therefore have found a beginning at an earlier period—and perhaps all the first efforts have perished. Modern jewellers work with a lens, and with its help they discover imperfections not visible to the naked eye, or observable by the home connoisseur who is less familiar with the technicalities of the craft. The difference between hand work and machine-made goods is, however, easily understood, and does not require an expert to point out. Quite an ordinary collector will note the suggestion of original work seen in all hand-made goods where tools have been given latitude and the operator has imparted something of his own personality to the object he fashioned.

The maker of cheap jewellery aims at effect rather than quality of workmanship, whereas the older worker preferred to merit the approval of his patron for quality and stability rather than for appearance. The old worker would take the gem in his hand and work round it a suitable setting, he would labour to give the jewel the setting best adapted to its size, shape and quality. Many modern artists, however, in contrast take a framework of gold or silver shaped according to a standard pattern, and select a stone the nearest they can to fill the setting, or to cover up the shoddy workmanship which needs something to hide the defects or the loose way in which joints and frames are made and held together.

The setting of gems is referred to at some length in another chapter, suffice it to say here that the gem should be set in harmony with colour, size, beauty and lustre; and when the golden frame is the first consideration, which it must often be in the making of important works for special purposes, then the stones selected for its enrichment ought to be of the most suitable obtainable, and the matching complete; even then the gems to suit the design ought to be chosen before the setting is formed, for stones are not things to be cut and spoiled in order to make "them fit."

For many long years the work of the jeweller was the exposition of the man, of the interpretation he conceived of his art, and of his personal views of what was proper and fitting. The ideals of the craftsman showed clearly in the handicraft he followed, and especially in the work he accomplished. It was a craft in which great individuality was observable, although in quite early jewellery there was a well established rule of form and size. Take, for instance, the *fibulæ* of the Romans; in a small collection there is much sameness, but although there is a well defined pattern the worker was allowed full freedom in his interpretation of the model, and as many of the tools he used were of his own fashioning the articles he produced by their use would be slightly different from those of others who would use tools varying in form, and giving different effects although used in a similar way. The engraver handles his tool according to the formation of the graver, and he uses certain tools with which he is most familiar, and produces better results than if he used those of different forms or sizes; in this way then some of the technicalities of craftsmanship are explained.

Modern machinery has altered the results secured by simple tools worked by the craftsman who exercised his judgment as to the way he worked them. For years past the technicalities of trade have been growing narrower, and the use of machinery and modern tools of standard patterns and weights has brought about standardised goods, robbed the craftsman of much of his freedom, and lessened the value of the artist who one time free must now, except under rare conditions, follow the lines laid down by his employers.

COMMON PRACTICE.

What is called common practice in trade is that guiding principle which insists upon a common basis of production, following an established rule. This is seen the more clearly in the present-day, but it existed nevertheless in olden time.

In Chapter V., "*Guilds, and the Influence they Exercised*," it is shown that the guilds exercised a strong influence upon the workman and ensured his adherence to "common practice" in the past as to-day. Indeed the goldsmiths and others were bound down by regulations in the past even more than individual makers and their employés are now.

It may be thought that this is not a matter of much importance to the collector. It is, however, in that its realisation leads to understanding the similarity between many of the trinkets and jewellery made in very remote periods, and also helps to make us understand how it is that there was never any very great divergence between the makers throughout long periods of time. The making of jewellery is but the story of the evolution of art, for there are very few striking novelties or original departures from the common practice of the day. The difference lies in the quality of finish, in the freedom allowed in decoration, the difference in the metals used, and the way in which they were manipulated.

The common practice is seen in barbaric jewellery, in the arts of early peoples of all ages, and in the art of those who worked for uncivilised races. Many of the beautiful jewels of Oriental peoples, very crude in their formation, and made, possibly, to impart a sense of splendour, an attribute of Oriental pride, were fashioned with that object in view. It does not follow, however, that the workmanship of all the jewels which have been preserved or that were made at any one time, and by or for any one people, was the very best the workman could turn out. To-day the workshops of Birmingham and other manufacturing places do not produce all their goods of one standard quality. Far from it, as art advances there are more patterns and a greater variety of wares to suit the tastes of everybody and to fit their pockets too. Thus in line with the markets catered for, the maker produces his wares; he tries rather to please the tastes of his customers than to follow his own ideals, and he works accordingly.

When art passed out of the hands of the amateur it went into the hands of a maker who, for money or as in earlier times barter, produced goods according to the buyer's needs. It is a well established fact that in the days of the Stone Age there were workers in flints who supplied many fighters and hunters, and even the needs of a settlement in domestic flints. Entire workshops, and the refuse such workings would produce, have been discovered. If there were craftsmen and traders in the limited supplies of those far-off days then we may be sure when jewellery was first worn the amateur would soon surrender his occupation to those who by practice or skill were better able to carry it on. The achievement of the trade under the guidance of their guilds, and the results secured by some of the best artists who have worked in metals and in the cutting and setting of gems are of considerable interest; the common practice of the earlier times is, however, of more interest to the collector, in that the collector who understands something of the methods of production of the objects he admires and collects is better able to appreciate his treasures, and far better able to secure bargains, for he is independent of his agent and able to assess the value of his curios from the standpoint of the worker and artist, as well as by their scarcity and rareness and their curio worth.

There have been times when production has been very prolific, and there have been times when it would have been impossible for much wealth in jewels to be accumulated. The peoples in some countries had access to materials which were denied to others. We have heard of the plentiful supplies of pure gold which once were to be had by a few simple mining operations in this country; and in some parts of Central Africa the natives, although perhaps possessing no other wealth, had all the gold they required to make for themselves bracelets and armlets and rings of the pure metal, the material costing little, and the workmanship a mere trifle, if the wearers were not the actual makers of those objects which they were so proud to wear.

AMATEUR REPAIRS.

The home connoisseur is not an admirer of modern art when it is far removed from the antique. The collector from the very bent of his mind and "the place wherein his heart is set" must have some preference for "old style," and often for that which is farthest removed from his own modern surroundings. As will be seen in other parts of this work the collector of jewellery, and those who delight in the contents of the old jewel box which has come down to them as a heirloom, must of necessity be better acquainted with Georgian or early Victorian jewellery than with ancient Roman or Celtic art. When, however, the amateur worker in precious metals and fashioner of jewellery begins to copy the antique he generally takes as his model the objects which were made by man in the days before he was very far advanced from amateur home working. Many of the oldest works of art are really very beautiful and can be copied without difficulty by the present-day amateur; and what to the collector is, perhaps, of more importance, they can be repaired when broken.

A knowledge of craftsmanship, even if only that of an amateur worker in metals is useful to the collector in the pursuit of his hobby. It enables him to buy oddments with some slight blemish which can be repaired without difficulty by those who have learned the use of simple tools. To take such repairs to a professional is to ensure their repair on the lines of modern craftsmanship, which almost invariably shows their restoration.

The amateur repairer follows the lines of the older and less skilled man who was content with a cruder finish. He is careful not to remove the marks of age, but lovingly deals with such indications; in short he combines the skill of an amateur craftsman with the veneration of the antiquarian. Many bargains have been acquired in this way and not a few choice objects have been secured in a damaged state, and then repaired; in olden time and at the hands of Eastern workers rare jewels were often given indifferent settings, causing them to be put on one side as damaged after a few years wear.

Some very remarkable jewellery has come to us from the East; and among the barbaric jewellery of nations outside the reach of modern machine-using peoples there is still much interesting native art. Races which were but a short time ago in a state of savagery possess beautiful jewels which they set in gold and other metals in primitive styles, and at the auction room such objects—needing only a little repair—often change hands at prices which should satisfy the most economic collector.

The older jewellery is hammered; wrought by "hammer and hand," and the wires by which it is sometimes linked were often beaten together. The smith of early days had strong faith in the value of welding, and he laboured to produce a true weld without the use of modern tools, drills, rivets and screws. The jeweller in like manner operated the more costly metals and achieved his more delicate craft in a similar way.

Some of the old jewellery is remarkable for its great simplicity; necklaces were formed of delicate hollow cubes of gold, beads of stone and pottery, glass, and perhaps a few pearls, the entire ornament, gracefully fashioned, having a simple wire loop or hook by which it was fastened.

The study of a few objects of contemporary art makes it easy to form the right ornament which may be missing. It may be a lozenge or a cube or some other elementary object, for necklaces and strings of beads were seldom formed of the same ornament often repeated. There was variety of shape, but always symmetrical in the artist's conception of primitive beauty; thus it is that what is sometimes termed "natural" art is much admired—and rightly too when it is remembered that simple objects shaped by man in early days were copies, although perhaps crude, of models provided by Nature, and between Nature and Art there is a close affinity.

Bead necklaces are very often in need of repair. Fine strong silk is best for threading beads, although hair is sometimes employed, and occasionally fine wire, but the latter is not flexible enough for beads which are to be worn round the neck. A bead needle is an instrument easily procured and is useful for the purpose, as it will thread the finest pearls or seed beads—it is long and thin being the same thickness all its length. Most bead necklaces are fastened by small gold clasps, which in many of the older necklaces are much worn. It is not worth while trying to repair these, or having them repaired by a workman; modern clasps are on the same pattern and can be obtained almost identical with those made a century or more ago.

When repairing bracelets and supplying parts missing the

amateur is apt to forget the usual sizes, and liable to make them too large.

Brooches are often out of repair, but a new pin will generally make them "as good as new." Pins ready fashioned with and without plates with hinge complete can be bought from a working jeweller, and may be attached by the amateur who will leave the work not quite as new looking as the usual repairer of such things, but wearable. The decorative ornament of antique brooches frequently needs a little touching up to make the article presentable—and so the story of much needed repairs which a capable amateur can carry out could be extended. When once the collector learns that it is comparatively easy to keep old jewellery in fairly good order the owner—he or she—is not likely to be satisfied with imperfect articles to wear, or to hand on to posterity as relics of the past!

SIMPLE TOOLS.

The tools required by the amateur are those which were used by the early artists who fashioned the antique jewellery under repair, and they are similarly made. That of course refers to the later works of antiquity and not to prehistoric objects when the patience of the workman must have been sorely tried, although perhaps he had not learned the value or measurement of time!

The tools procurable now are no doubt better in form and

finish than those used in olden time, for they are fashioned by machinery and turned out in quantity. In the hands of the amateur, however, these better tools will not enable him to turn out better work than the makers of the jewellery undergoing repair, they may, however, be some little compensation for the shortage of experience. One or two hammers are essential for operating the chasing tools and the gravers, which can be bought in a variety of forms, and slightly different in shapes. Files are very useful, and several of the makers of amateurs' tools are selling very handy sets of small files. A bench stake is necessary—it can be fitted into an old table and removed at will. Shears for cutting the metal are needed, a few chisels and a drill for fine work. A jeweller's saw frame and saws will come in handy, and of course a sand bag, the latter to hold the work in place. The amateur metal worker should take a few lessons, for the workshop practice of a working jeweller is much too big a subject to be handled in this chapter.

A little sheet silver and some wire are materials needed. Gold or amalgam wire will be helpful in repairing chains and links which a pair of plyers will shape to supply deficiencies. Incidentally it may be mentioned that repaired parts can be burnished to the condition of the old and a good polish given where needed by the use of Tripoli powder. In resetting old stones or when matching them a small piece of tin-foil at the back acts as a reflector, some use black paper behind opals, as will be seen when examining the setting of an old piece when it is necessary to repair the frame of the stone or to fit a new stone in the old setting. Care should be taken to fit the stone in tight, that is to say inlay it, and then with a small burnisher or similar tool smooth over the edges until they hold the stone firmly. If such edges exist and the stone has to be put into the old frame, then open out the rim until the stone is put in tight, and packed if needed.

THE RESULT.

The result of a smattering of knowledge of craftsmanship, as gained from careful observation and from actual practice is that in a collection of old jewellery the best effects will be secured. There will be no imperfect and meaningless pieces, the real use of which is not observable because parts are missing; and there will be no necklaces without clasps or bracelets unstrung or partly defective. There is much to interest in the possession of the antique, but perfect specimens in every branch of collection are aimed at. In jewellery, however, it has been shown that it is not always possible to secure perfect examples, and that when bargains are going it is because there is something deficient, perhaps only a trifle which the art of the amateur craftsman can put right.

GUILDS, AND THE INFLUENCE THEY EXERCISED.

LONDON GUILDS—SCOTCH AND IRISH GUILDS—MEN OF MARK—SOME RETAIL JEWELLERS.

THERE is something very clannish about workmen, a kinship and fellow feeling which has been apparent in all ages. There are few records extant, it is true, but there is abundant evidence that in very early days workers in the same metals, and those who produced similar objects, consorted together, and often dwelt in the same streets and thoroughfares for protection, and, as trade developed, for the better pursuance of their business. This fellow sympathy between men practising the same arts is seen in almost every trade. At the dawn of the Christian era the silversmiths of Damascus had a common cause, and a leader in Demetrius. The workers in gold and silver have generally been among the most prominent craftsmen, and that is as it should be, for the metals from the very earliest times represented the wealth of nations, not only in bullion and coined money but in jewellery and plate.

In another chapter the beauty and clever working of gold in prehistoric times, and in the days when the world was young, is more fully dealt with, the connoisseur must, however, remember that the precious metals in some form or other represented the world's wealth, and that those possessions which were in kind were made attractive by the genius of man, who gained proficiency in his craft as the knowledge of the art became better known. The furtherance of that object and the retention of the skill acquired, together with a growing desire to conserve the trade of a town or district brought about the formation of guilds—a form of trade protection which in time exerted such a widespread influence upon craftsmanship and art.

Such guilds as those connected with the art of the goldsmith and the jeweller have been met with in many countries, and some of them can trace their origin to very early times. In England there are many old guilds, but most of them have now lost much of their original value, although in more recent years there has been an attempt to revive guilds, federations and associations of traders for mutual protection, and in some instances for the better and purer practice of the craft with which they are associated.

LONDON GUILDS.

It is not surprising that among the numerous guilds which have been formed in the metropolis the workers in precious metals have played a prominent part. Very early the workers in gold banded themselves together, and the Goldsmiths' Company soon became one of the wealthiest and most important guilds. The Worshipful Company of Goldsmiths was no sinecure, for it acted beneficially upon the craft and instituted a system of purity of the metal used and fixed a standard by which the quality of the wares wrought by the members could be gauged and known.

The same Company has like a few others retained its beneficial influence on the trade, and still at its Hall marks the plate produced within its jurisdiction, thus performing a useful function in the State, taking part in the regulation of its trade and commerce. It may be mentioned that the purity of the coin of this realm has also been under the control of the Goldsmiths' Company, for the historic trial of the Pyx is one of its functions, and the assay of the currency to ascertain its quality is entrusted to it.

The records of the earliest happenings of the Guild are lost, but in its magnificent Hall there are many rare relics of its former usefulness, of the works its members produced, and of the more modern treasures in the making of which the goldsmiths of recent times have well maintained the high reputation which their predecessors in the craft enjoyed. The Company was active as far back as the twelfth century, and like many of the older guilds was of a semi-religious order, mainly caring for the maintenance of the standard of quality of materials used and of the work turned out. The Goldsmiths were craftsmen, and cunning workers they were too, as the rare pieces of early make still extant show. The ancient mystery had its patron saint, and St. Dunstan was honoured in that he was himself a worker of metals. There is a legend that Edward I. possessed a ring of gold in which was set a famous sapphire, the ring being the handiwork of the Saint. St. Dunstan's day was formerly kept as a gala day by the members, and bells were rung and prayers said for the souls of deceased members of the craft.

According to ancient charter Edward III. granted the goldsmiths special privileges, and ordered that the Company should exercise an oversight over all the goldsmiths, most of whom had their shops in the High Street of Chepe. Perhaps one of the most important duties associated with the manufacture of jewellery in those days was the thorough way in which the Company scrutinised all the articles which had then to be sold either in Cheapside or in the Exchange. Cheap jewellery was made even then, for we are told that the practice was to cover tin with gold so cleverly that it was easy to deceive the public and to palm off false goods, and in addition to cheapen the production of jewels by using counterfeit stones.

It is interesting to note that the first Hall of this powerful guild was erected in 1350, a more important one followed, but alas! it was destroyed in the Great Fire. The present magnificent Hall on the original site was not built until 1835, it is therefore comparatively modern, and is very handsome and decorative. It contains much plate and many objects of great value which will be more fully referred to in a future volume of this series dealing with gold and silver plate, for jewellery is but a minor part of the work of the members of the craft.

Among the smaller companies in London the Girdlers' Company claims to be one of the oldest. It carries us back to the days when girdles of silk were worn, and to those times before pockets were in common use. The origin of this Company is said to be found in some lay brethren of the Order of St. Lawrence supporting themselves by the manufacture of girdles. The guild, however, dates from the time of Edward III., and it gained many subsequent charters. In the reign of Elizabeth the Girdlers joined the Pinners and Wyreworkers. For many years they prospered, and made many jewelled girdles and supplied those charming belts from which were suspended so many oddments, and which served to keep close at hand chatelaine instruments and the numerous trinkets which the orderly and careful housewife thought necessary to possess and carry. (*See* Chapter XXVII., "*Chatelaines, Chains and Pendants.*")

SCOTCH AND IRISH GUILDS.

Some interesting particulars about old Scotch and Irish plate are given in Mr. Cripps' standard work on *Old English*

Plate, but those metal workers were chiefly devoted to the manufacture of plate as separate from jewellery. The trade as a whole was conserved with the same care in Scotland and Ireland as in England, and the Guilds established in the two countries acted in their respective spheres much as did the Goldsmiths' Company in England. Mr. Cripps says "Then came the letters patent of King James VI., granted in 1596, and ratified by parliament in the following year, to the deacon and masters of the Goldsmiths' craft in Edinburgh, which gave further effect to these statutes by empowering that body to search for gold and silver work, and to try whether it were of the fineness required by law, and to seize all that should appear deficient; this gave them a monopoly of their trade and the entire regulation of it, separating them finally from all association with the 'hammermen' or common smiths." The assay office was then in Edinburgh, the Glasgow office being established at a much later date.

The Irish goldsmiths were banded together under the title of the Goldsmiths of Dublin, and had duties to perform in order to keep up the standard and quality of the work made in Ireland, at any rate within their immediate jurisdiction. Mr. Cripps mentions a Company of Goldsmiths in Cork who "marked their plate with a galleon and a castle with a flagstaff."

Scotch jewellery has always had characteristic symbolic designs, and the stones which have been used, together with the jewellery representing the clans, have given the gold and silver work of the northern part of Great Britain a distinctive style. Ireland, however, has during modern times shown no particular preference, other perhaps than the form of the Irish harp, which together with the shamrock incorporated in modern jewellery has distinguished it; it is in the early jewellery of Ireland that collectors find the greatest interest. In Chapter X., "*Celtic Gold*," several pieces of this quaint and rare gold and silver work are described. In the making of this, however, no guild exercised any control, although in the very earliest forms there does appear to have been concerted action, and a definite plan of ornamentation and form carefully carried out by those early workers in the precious metals, although they may have unconsciously followed their own bent, which was narrowed by the limited knowledge of art as then understood, and in its limitations preserved the purity of style and of metals used without any formal recognition of control.

It should be mentioned too, that foreign trade influence has at all times exercised considerable control over craftsmanship in this country, and in all the trade and commerce which have brought to our doors commodities, especially those in which art is seen.

MEN OF MARK.

The functions of the goldsmiths of London became more extensive as time went on. These traders lent money and bullion to the King and nation, and in many instances provided for the country's needs in time of war. The Jews too, in early days were great traders and financiers. They congregated in Old Jewry which still retains the name of their location; they were, however, banished in the thirteenth century. Then came the Lombards, who settled in what is now Lombard Street, and there they hung out their signs and traded in gold, and became merchants, importing goods from foreign parts, eventually founding those great banking houses which have played such an important part in the nation's finance.

The goldsmiths of Lombard Street had among their number many men of mark. It was there that Sir Thomas Gresham traded at the sign of the "Grasshopper"; and when he built the first Exchange, thus providing merchants with a meeting place, and consolidating the commerce of the country, earned the esteem of his fellow citizens. In the same thoroughfare lived Sir Martin Bowes, a goldsmith, in the reign of Henry VIII.; and at the sign of the "Unicorn" was the shop of Edward Blackwell. It is recorded too, that Sir Robert Vyner carried on the business of a goldsmith in Lombard Street, and he it was who made the new crown for Charles II., after the Restoration.

Many of these old goldsmiths, and in some smaller degree jewellers, were broadminded men who gave much of their time to public affairs and sought in one way or another to improve their city. One of these men of note was Richard Myddleton, whose name is still perpetuated in Myddleton Square, Clerkenwell, near the site of the culmination of the great achievement of his public career. Born in Denbigh Richard Myddleton became apprenticed to the Goldsmiths' Company, and in due time commenced business on his own account in London. He seems to have been a very successful trader and craftsman, for in 1597 he represented his native town, Denbigh, in Parliament. Royal patronage was of great value then, and indeed carried with it support and often financial aid. King James I. made him "royal jeweller," and often visited his shop. When Myddleton evolved his great scheme for supplying the Metropolis with pure water from Hertfordshire the King joined in the enterprise. It was successful, and in course of time the New River brought a plentiful supply to the New River Head, then in the fields between Islington and London.

The goldsmiths of the past continued to trade in smaller things, and many had small shops, although in a quiet way they were lending money and founding the great banking houses, some of which are yet extant. The work of these goldsmiths is not likely to be met with among the jewellery of the "home connoisseur," but these men of note are worthy of veneration, for they showed how in those days it was possible for simple craftsmen to serve their country as artists of a rare order, as well as building up great businesses as merchants, which were surely, if slowly, laying the foundations of Britain's trade all over the world, and forging the golden chain of commerce which has been her bulwark, and enabled her to resist her enemies and hold her own as she moved on in the growth of empire.

SOME RETAIL JEWELLERS.

The craft of more modern days seems to have been divided between those who made jewellery and sold it in their shops, and those who were content with buying from others and retailing it. The business of the present day is dual too, although the larger manufacturing concerns are distinct from the shops in which jewellery of various kinds is offered to the public. It is not easy to disassociate the craftsman from the thing he has created when his personality is known, and it was especially so in olden time when each object was stamped with the impress of the maker, and often bore marks of his individuality. In the following chapters, however, for various reasons, the objects of art representing the crafts of many nations, and created at different periods, must be reviewed either according to the period and style in which they were made, or the particular object or purpose of the jewellery, and not with reference to the artist or retailer. We may, however, indulge for a moment in picturing the conditions under which some of the old jewellery was sold in the past.

Cheapside was the chief market—"Chepe" of olden time—and it was in Goldsmiths' Row that the goldsmiths plied their trade, in like manner the silversmiths worked and sold their wares in Silver Street, near the market of "Chepe." As we have seen the Lombards and goldsmiths hung out their signs, and some adopted the sign of the "three golden balls" still associated with those who combine the retailing of jewellery and the lending of money. The old taverns, often connected with traders' shops, not infrequently adopted emblems of the craftsmen by whom they were chiefly supported. A study of the old trade stationery of retailers of more than a hundred years ago often throws some light upon the way in which they did business, and also tells of their patrons; and their bills, perchance, tell of their customers and of the goods they sold to them.

The "Golden Angel," alone or in combination, was a common sign in the eighteenth century. It was adopted by Ellis Gamble, a goldsmith of Cranbourn Alley, who was evidently proud of the sign, and had a very large trade card (*see* Figure 2) on which it appears—it is an unusually fine card, and records the trader's status, and the work he carried on. Ellis Gamble of the "Golden Angel" entered his mark at the Goldsmiths' Hall in 1696. He appears to have been a silversmith of some notoriety too, and was the son of William Gamble, who had traded in Foster Lane at an earlier date. It was to Ellis Gamble that in 1712 Hogarth was apprenticed, and there learned the art of engraving metal. Hogarth in after years engraved several trade cards, one being for his former employer. Gamble, according to his card, made, bought and sold "all sorts of plate, rings and jewels."

The old styles of engraving and of the designs then prevalent in decoration are reflected on many of the trade cards and in the catalogues of traders. The "Angel" was a favourite sign, for Smith, in "ye Great Old Bailey," who had a pretty card in "Chippendale" style, used it, as well as others. Chalmers and Robinson were jewellers at the sign of the "Golden Spectacles" in Sidney's Alley, their cards were in "Chippendale" and pictorial styles. At the sign of the "Golden Ball" in Panton Street, Johnston & Geddes were said to "sell all sorts of jewellers' work."

The bill head, shown in Figure 3, is doubly interesting, for not only does it represent a quaint style of stationery, on which is exhibited on a shield of arms of the period the style and name of the retailer—T. Hawley—but it gives the details of the whereabouts in Strand where the shop was situated, "three doors from the Adelphi." This is the original invoice of a gold watch made for the Duke of Wellington (then Marquis of Wellington) in 1813, two years before the battle of Waterloo—perchance the watch he wore on that great day.

The description of the watch reads "Small gold watch and gold key, with the engraving of the Marquis of Wellington." The price named is £8 3*s.* 0*d.* Wonderfully interesting are these old trade cards and shop bills of a century or more ago!

THE ENGRAVER

THE MANNER OF ORNAMENTATION—HISTORY OF THE ART—SOME EXAMPLES—TECHNICAL POINTS.

IT is well to be acquainted with the different artists by whom objects of art are worked. The engraver is one who practices the art of ornamentation and generally works upon some object already fashioned. His scheme of ornamentation is dependent upon the article he embellishes, and he is frequently compelled to work according to the purpose for which the object is to be used; moreover he often finds it necessary to adapt some given style of pattern so as to make it conform to the shape or size of the piece of metal or other material upon which he operates. The art as applied to the jewellery trade is of course only one branch of engraving; it is the decorative side, and includes that of fanciful ornament and rigid pattern, also embracing the engraving of legends and inscriptions, especially so in the case of presentation ornaments. It is applied to a much larger extent upon gold and silver plate which is so often engraved; but trinkets—like snuff-boxes—have often been engraved with inscriptions and short sentimental sentences, and in some instances the engraver has almost covered the surface of the object with monograms and names. The art as applied to metal is very ancient, and although much of the so-called engraving is more in the form of chasing, the graver's tool has long been known and applied for the adornment of jewellery.

THE MANNER OF ORNAMENTATION.

The manner of ornamenting and embellishing jewellery does not trouble the collector of old jewellery overmuch—he rather judges the effect of the craftsmanship than the methods by which the results have been achieved. In order to fully appreciate the finished article it is, however, well to understand something about the technique of ornament. Moulding, casting, hammering and putting into shape are some of the first processes by which the maker of old jewellery produced the rough object. Today machinery plays an important part not only in producing the article, but in finishing off the details and finally polishing, and if necessary cutting and engraving the ornament. The hand work which meant much time and labour is now only reserved for the more costly work, and for inscriptions and minute strokes of the graver. In olden time before machines in which such delicate things as small pieces of jewellery could be operated were known, the artist fondly handled the article and gradually by graver and other tool added little by little to the decoration, until a finished article lay on his table—made, engraved and polished by hand.

Some of the methods of craftsmanship have already been referred to, and some of them have been employed without interruption from the very earliest times. Among these the engraver, working with quite simple tools, has been the most prominent artist.

The engraver, as it has been stated, in more recent times has worked almost independently of the jeweller, and has carried on his work at the beck and call of the maker. His functions have been two-fold, for he has provided the engraved gem for its setting, as well as engraving the gold and silver setting and ornamenting the metal object which is frequently without stones or other additions. The gem engraved with signet or seal, as apart from the work of the lapidary and polisher, has contributed much to the beauty of the metal setting. In olden time, perhaps, the engravers of stones and gems made their setting, and it is well known that many skilful jewellers have made and engraved, and sometimes inscribed jewellery and fancy trinkets, carrying through the whole process. Now, as it has been stated, engraving is only one of the branches of applied metal art.

Although engaged at times upon fashionable replicas, and copying the styles of former periods, the engraver has generally followed the prevailing taste or style of the period in which he worked, and his work has taken the form of emblem, device or letters in the gold or other metal operated upon, or in decorating a plain surface and making it a picture in low relief, as distinct from the raised or moulded ornament in which so much of the old gold jewellery is so rich.

In some instances the effect of relief is much enhanced by the additional cutting to which the ring, brooch or other object has been subjected. Larger pieces of plate, especially silver plate, are frequently engraved with shield, monogram or inscription; and quite small objects are likewise engraved with legends and sentimental mottoes—as in posy rings. (*See* page 238.)

HISTORY OF THE ART.

From early examples the use of the graver (the tool by which engraving is effected) was known to the ancients who operated on precious metals as well as on stones and gems. How far back this art was known it is now difficult to ascertain, writers are usually content with Biblical proof, and about the use of the graver the story of the Jews is quite clear. An often repeated reference is that mentioned in Exodus of the plate of pure gold on which was engraved "Holy to the Lord," that being an instance of the engraving of letters or characters, the plate being still further enriched by ornament. In this early mention of engraving the art is brought down to the equally early mention of jewellery and gems of which the setting of precious stones in Aaron's breastplate is sufficient evidence. It is very interesting to know from such an authentic source that the craft was followed by skilled workmen, the name of one at least being recorded. In the furnishing of the Temple the Israelites called to their aid men skilled in all the crafts; and in Exodus special mention is made of Bezalel, who appears to have been a master craftsman, well skilled and cunning "to work in gold, and in silver and brass, and in the cutting of stones for setting." In this work he was associated with Choloab who was evidently another ancient Jewish worker in the precious metals and gems.

It may be assumed that the Jews learned this art of making jewellery and engraving it from the Egyptians, whose still earlier work is evidenced by the discoveries in their tombs, and confirmed by Scriptural mention of the treasures given to the Israelites by the Egyptians when they "asked of the Egyptians jewels of silver and jewels of gold. . . . And they spoiled the Egyptians."

The graver's tool was well understood in the land of the Israelites, so much so that it was used as a symbol of higher things. In Zechariah iii., 9, we read, "For behold the stone that I have set before Joshua: upon one stone are seven eyes, behold I will engrave the graving thereof, saith the Lord of Hosts."

The wealth of gold vessels and scarce objects made for the Jewish ceremonial cannot be estimated from the brief records given in Holy Writ—all these priceless treasures have long since gone, and the world is the poorer. What would a collector give for a piece of ancient Jewish jewellery? The worship of the God of the Hebrews involved no burial of relics—but contemporary and even earlier Egyptian tombs have yielded treasures of this art as then practised, and we must be content with them as mementoes of those days when the East was in the ascendant, and the Western world practically unknown.

SOME EXAMPLES.

The Egyptian rooms of the British Museum are full of examples of early engravings. Gold rings are covered with inscriptions, most of the bezels being symbolical, and some cleverly executed although very minute: to take an instance there is one on which is a man-headed lion crushing a prostrate foe with his paw, on the other side being an inscription which means "Beautiful god, conqueror of all lands, Men-kheper-Ra." On another ring, on the bezel, is the figure of a goddess seated in a boat under a canopy. (For further mention of inscribed Egyptian rings and jewellery, *see* Chapter VIII., "*Egyptian and Assyrian Jewellery.*")

The large amount of Celtic gold jewellery found in Ireland shows that British goldsmiths were no mean craftsmen. There is little to tell in what way the early British benefitted by their close touch with Roman craftsmen during the first four centuries of the Christian era, though there is evidence that when left to their own resources they still pursued the art—for Briton and Roman were merged in one nation.

In Saxon times jewellers continued to work in the precious metals, but few personal relics have been preserved to us—and few indeed are the inscribed jewels from which their owners are known. One rare example, often quoted and described, is the famous relic of Alfred the Great now in safe custody in the Ashmolean Museum at Oxford. This rare treasure was taken by the King on his retreat to the Isle of Athelney where it was found. It appears to be crudely fashioned but in reality exhibits considerable skill in manufacture and in the art of engraving. (*See* figure 10). (For further reference, *see* Chapter XII., "*Anglo-Saxon Gold and Silver.*")

TECHNICAL POINTS.

Crest, monogram and inscription are common enough on domestic plate, and in some instances they occur on jewellery and trinkets. As a familiar instance snuff-boxes may be mentioned. The engraver has often cut monograms on signet rings and in more modern days worked elaborate monograms on lockets and pendants. Coats of arms are of course more restricted in their use, but they are not infrequently found on old trinkets which have changed hands many times, now finding a home in the possession of those who are in no way associated with the arms or crests engraved upon them, or persons who are now entitled to use or exhibit them. It is always interesting to understand the various engravers' marks which in heraldry mean so much. A knowledge of some of these may be useful, although the great differences in crests and symbols made by a few strokes of the graving tool cannot be given here at any length. Colour in heraldry is important, denoting different blazonry. The principal colours and their heraldic names are as follows:—gold (*or*), also denoted by yellow; silver (*argent*); black (*sable*); blue (*azure*); red (*gules*); green (*vert*); and purple (*purpure*).

In the sixteenth century the engraver on metal began to denote the heraldic colours of his patron's shield by lines and dots, which became the method of denoting colours then generally accepted. Taking these in the order already mentioned, the dots and lines used, which may be clearly seen on heraldic engraving on curios and trinkets are as follows:—gold, *dots;* silver, *plain ground;* black, *crossed vertical and horizontal lines;* blue, *horizontal lines;* red, *perpendicular lines;* green, *lines from right to left*, and purple, *lines from left to right*. By using these key notes the true colours of the common armorial bearings and shields can be ascertained.

SPECIAL LAPIDARY TECHNIC

Sphere Cutting

ALFRED M. KRAMM

During the middle of last year, after eyeing a one pound hunk of crystal-clear glass, which the writer had purchased several years before and after reading an article on sphere cutting, he could not resist any longer the urge to try his own hand at it.

In cutting this sphere, (a) represents the first 4 1/2 pounds removed by diamond sawing, (b) a pile of chips from saw "chipping," (c) finer fragments removed by sawing and grinding, (d) the final sphere on a pedestal mount for display.

Knowing nothing at all on the subject, except by reading on it now and then, it was thought advisable to carefully re-read any such article before proceeding with the actual work. Knowing what constitutes a perfect sphere, it was aimed at producing one if at all possible, as good as the best. Naturally, all printed instructions were followed with great care. A pretty good looking sphere was the result of 32 hours of hard work. It measured 2 1/2 inches in its diameter. During its process of completion a number of unwelcome scratches showed up which had to be polished out.

This not only consumed extra time but also tended to reduce the ultimate size of the sphere beyond its originally contemplated measurement. By correcting many or some of these weaknesses, it was thought possible to not only speed up the making of spheres, but also to make the hobby a more pleasant one.

It did not seem possible that even the most expert lapidary could take a perfectly fine ground and optically true sphere from the sphere-cutting tool to the sanding drum and later to the felt polishing wheel, without disturbing in the least its former optically true surface.

Lenses, which are segments of spheres, are not made that way, neither can whole spheres be made that way and be perfect. It was believed that there must be a method between the two.

All of these thoughts ended up in the purchase of another chunk of optical glass which was to be large enough to produce a 4-inch sphere, in the making of which the writer was willing to try any short-cut or unconventional method as long as it looked promising to do a *better* job *faster*.

Unfortunately the chunk of glass proved to be badly off shape. It weighed 9 pounds to start with and was reduced to 2 1/2 pounds, which was the final weight of the 3.75-inch sphere.

Cutting this block into a near cube and thereafter removing its 8 corners by sawing, cut its original weight down to exactly one-half. So far, so good.

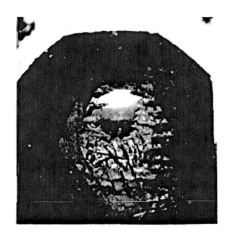

FIG. 1.

A lot has been said and put in print about the trouble one gets into when he trys to grind down a rock with a grinding wheel that weighs little or no more than the rock itself. This is what "yours truly" tried to do. Experience is a good teacher. In less than 15 minutes of grinding the wheel was so bumpy that a continuation of it would have surely damaged either one or the other, if not both. The practical lesson has not been forgotten. It was plainly a matter of changing methods then and there, or go back to marble grinding.

But necessity is the mother of invention and marble grinding would not solve the problem. Somehow, the extreme unevenness of the partially rounded block had to be cut down considerably before the sphere cutting attachment should be used, even if all previous rules concerning the art of grinding rock, especially glass, had to be broken.

So the diamond saw blade, with which the block of glass had been cut so far, was removed from its arbor and replaced by one of 6 others that had been bought for 50 cents each. While this blade was not so hot, that is, out of true in two directions, it never-the-less cut. It was contemplated to run it dry against the sphere, ripping inch-long shallow corrugations into its high spots and by applying some extra pressure, develop enough heat within the saw to have it *do* as much chipping as it would cutting. The bumpiness of the blade was somewhat neutralized by holding the sphere against it (by hand) at a slight angle.

The idea worked. Soon after a few grooves had been cut and the saw became hot small flakes of glass began to fly. Greater pressure often would produce chips as large as 1/2-inch. Soon a special technic was developed by which the saw-chipping was done by circling a given point, closing in on its center, until the last of it dropped off like an ugly wart.

FIG. 2.

After 1 1/2 hours of such treatment or a total of 5 1/4 hours from the start, the 9-pound hunk of glass had been reduced to 3 3/4 pounds. At this stage, photo Fig. 1 was taken. In it one can still see many of its original saw cuts, plus two clear spots, opposite to each other, through which distant scenery may be viewed.

It was at this point when the home-made sphere cutting device was set up for action. (See Fig. 5.) This tool represents nothing new in itself. It might be added though that the tool, which was about 6 inches high, and could have been an inch or so lower, was filled on its inside with rags and these were covered with melted asphalt up to about 3/8-inch from the top. The remaining space was filled with desired abrasive which in this case was 100 grit in the start.

At first this tool was used but for a few minutes at a time, that is just long enough to indicate the various high spots on the sphere which could be and were so much faster removed (see Fig. 2) by saw-chipping.

Gradually it became more and more difficult to see with certainty just where to continue cutting away and where not to. It was then when the whole sphere was coated with a quick drying black varnish, which the tool would remove only in its high points, and do it immediately, thereby indicating where further saw-cutting could be used to advantage and where it should not be used at all. In other words, this method did away with all guess work and made low cutting practically impossible. (See dark paint Fig. 2.)

FIG. 6

As the many and large flats on the sphere decreased in size and number, the saw-grinding was proportionally cut down, letting the tool grinding do the increasingly greater part of the work.

It had been noticed during this see-saw method of cutting, that is, the roughening of the sphere's surface by sawing, helped the subsequent tool-grinding in that it furnished many little lodging places for the grit, serving as supply reservoirs, which greatly contributed to the delaying of the ever-growing grip between the sphere and the tool.

After 11 hours and 42 minutes the sphere had now been cut

down in weight to exactly 3 pounds. In other words, NOW GET THIS, during less than 12 hours the original 9 pounds of glass had been reduced by 6 pounds, yet it took another 17 hours to remove the last ONE-HALF pound, which turned that hunk of glass into as good a sphere as the writer was able to produce.

At this point, that is after about 12 hours of careful labor, the largest diameter of the sphere was 4 inches and its smallest 3-13/16. Since at the start the sphere's diameter had been figured to be 4 inches, already 3/16-inch had been needlessly cut away. Another 1/16-inch was finally lost before the completion of of the job. Perhaps one with greater experience than that of the writer could have avoided, or at least reduced, such inaccuracy.

As the work progressed it became increasingly tiresome on the wrists to hold the sphere alternatingly in the right and left hand against the heavy pull of the tool.

After a total of 15 3/4 hours from the start, the last flat on the sphere had disappeared and the 100 grit, as used up to now, was changed to 150 grit.

The Craft of Jewellery Making

In 15 minutes this was replaced by 220 grit. No further change was made for the next 3 hours, during which period several scratches were inflicted upon the sphere that could not be accounted for, as in not over every half hour the tool, sphere and hands were carefully washed. Yet these scratches were nothing to worry over at this particular stage as they proved to be less deep than the thousands of remaining pits caused by the coarse grinding at an earlier stage.

After another half hour of grinding with 300 grit a complete clean-up was made and a small piece of wet canvas placed on the tool and the sphere set on top of it and pressed into the hollow below. To keep the canvas from sliding, a wire was wound (Fig. 5) around it. Twenty-five minutes of fine grinding with this arrangement, using 220 grit, really began to convince the writer that he was rushing matters. The semi-polish produced in this manner showed up large and small pits by the thousands. The canvas was removed and fine-grinding resumed on the hard base of the tool. This was continued for the next 3 hours, starting out with 300, and ending up with 400 grit. Now the sphere was really ready for polishing. So a new piece of canvas was fastened to a much cleansed tool and a very clean hand (and arms) placed the very clean sphere onto the tool (Picture No. 5), using for abrasive first tripoli and later rouge. This final process consumed another 5 hours before all of the minute pits had disappeared. The drag of the tool during these last few hours was at times terrific. But one thing was certain, with only reasonable care, while canvas was being used between the sphere cutter and the sphere itself, even when using 220 grit, scratches were out of the question. When a change of abrasive was needed a new piece of canvas was also put on.

The Craft of Jewellery Making

Any horizontal arbor can be fitted with sphere cups for grinding, sanding, and polishing spheres. This same basic machine can be made with a vertical shaft, a position many cutters prefer. One cup is mounted on the shaft, the other held by hand.

It is emphasized that all through the process no sandpaper was used, not even a felt or buffing wheel of any kind. The regular diamond charged saw, an old and partly worn-out saw, and a regular sphere cutting tool, first without and later with canvas, were the only tools used, excepting of course, an unsuccessful try-out for 15 minutes on an 8-inch grinding wheel.

The sleeve of the sphere cutter, on which the actual grinding and polishing was done, was not machined for the job, to run true, but ran itself true in short order. While the whole device was not well centered on the lap on which it was used, its eccentricity was always kept somewhat between 1/8 and 1/16 of an inch and, because of the slow motion of the lap plate, no trouble was experienced.

While the saw-grinding idea caused some little difficulty at first by sending fine chips of glass in all directions, including the face of the worker and his glasses, this trouble was later overcome by installing a piece of plate glass between the saw and his head. On a future job of this kind a shallow dish of water shall be placed directly under the saw, but not in contact with it, to catch most of the flying dust.

The sphere of 3.75 inches in diameter, as seen in picture No. 6, is extremely round with absolutely no flats or irregularities on it. This could not have been accomplished had it been sanded and polished by the use of the conventional sanding and polishing wheels.

In closing remember, it is of great importance that during the fine-grinding and polishing process the sphere must be continuously kept in motion. Never must it be allowed to remain at rest for as little as one second over the continuously revolving tool below it.

It is hoped that this somewhat novel method of sphere cutting will be tried out and reported upon by others. Thanks to *The Mineralogist* and other such magazines new ideas find their way into the many shops of amateur lapidaries. Speaking of himself, the writer has been repaid a hundred-fold for their small subscription costs, by reading these magazines from cover to cover.

Sphere Cutting Early Method

With the development of simplified technic the art of sphere cutting is again coming into popularity. For the past few years sphere cutting has become very popular amongst the southern California gem cutters.

Probably the first spheres to be cut were fashioned from crystal quartz, but just when this was first done has been lost in the antiquity of time. It is well known that sphere cutting started in the Orient many centuries ago, when the Chinese cut small spheres of quartz to be carried in the hand to cool the palms, a fashion which has continued to some extent down to the present time.

Spheres cut from quartz crystals have frequently been found in the ancient ruins, even of the Ninevites who used the spheres as burning lenses. In the time of Pliny, surgeons used crystal spheres for cauterizing by focusing the solar rays. Orpheus recommended that a crystal sphere be used to kindle sacrificial fires, thus ensuing the favor of the gods. The flame thus being kindled was called the fire of Vesta. At an early date seers made wide use of spheres and became known as "crystal gazers."

In recent years a good deal of sphere cutting has been carried on in old Mexico where a colorful variety of onyx (calcite) finds wide use. The Mexican spheres evidently are cut by crude methods for they are often "one sided." A perfectly cut sphere when rolled on a level sheet of glass should follow a reasonably true course over some distance. This, of course, is only an approximate test for true cutting.

Many Materials

By the aid of the recently developed technic spheres can be readily fashioned from any of the hard gem minerals. Large gemmy garnet crystals make superb spheres, and when the specimen shows asterism the finished sphere, if over two inches in diameter, may be rather valuable. In the Museum of Natural History at Cleveland, is a perfect garnet sphere some three inches in diameter which shows excellent asterism. When this specimen is rolled slowly the flashing "star" appears repeatedly at various positions on the surface. Almost any compact gem material of sufficient size can be utilized for sphere cutting.

The cutting of perfect spheres, contrary to general belief, is a relatively simple operation. Besides the ordinary lapidary saw, grinding wheels, and polishing buff, the only additional equipment required for sphere cutting is the special sphere cutting tool shown in the accompanying illustration. This sphere cutting tool can be operated either in the horizontal or vertical position. Special tools of this kind are available from various lapidary supply houses.

Cube First

If the material from which the sphere is to be cut is in the form of a large mass, it is then advisable to first saw out a cube. The corners of the cube are then rounded off on the grinding wheels until an approximate sphere is formed. The rough sphere is then made circular by the special sphere cutting tool and then polished in the usual manner on the felt buff.

Many rough specimens, like some agates and nodules, are already in the spherical form. The spherical agate filling a "thunder egg" can often be utilized to good advantage.

Sphere Cutting Tool

The sphere cutting tool consists essentially of various sized iron pipes which are attached to the shaft of an electric motor. Short lengths of ordinary iron pipe are excellent. The edge of the pipe should be slightly beveled. In the illustration it will be noted that the sphere is held in position while being rotated by a second piece of iron pipe held by hand. The short lengths of iron pipe are open at one end, and can thus be filled with the abrasive mixture. The diameter of the pipe should be slightly less than the diameter of the sphere desired.

Cutting is carried out by silicon carbide of various grits. The start is made with coarse grit, and final lapping is done with fine grits. This phase of the work is the same as preparing large flat surfaces for the polishing operation.

The horizontal running lap can be easily adapted for sphere cutting. The manufacturers supply sphere cutting tools that will fit or can be fitted to the horizontal lap. It will be seen that if sphere cutting is carried out with the tool in the upright position there will be many advantages. For one thing the lower reservoir (the lower pipe) can be filled with the abrasive and water mixture, and controlled better. However, good work can be done on either the horizontal lap unit or the equipment which is attached to the motor shaft and operated in the horizontal position. Special motors are available which can be operated in the vertical position, and in this case work can be done in the vertical position.

If you want to try sphere cutting without much special equipment, a cup can be mounted on almost any shaft. This cup is mounted on the top of a drill press spindle. The cup here is made from a standard pipe reducing nipple.

Polishing

After all deep scratches are removed from the sphere by lapping with very fine grit silicon carbide, polishing can be done in the usual manner on the felt buff. In polishing the ball can be held by hand and rotated frequently. It will be found that the curved surface of the sphere will polish quite readily. Curved surfaces are easier to polish than large flat surfaces.

Other Methods

There are various other methods by which spheres can be cut, and a number will be referred to here.

Commercially, spheres are usually cut on what is termed a "ball mill." The material is first rough ground to an approximate sphere and then a large number are placed in the ball mill. This equipment consists of two large circular plates, one above the other. The two plates rotate in opposite directions, and abrasive grit and water is fed in at intervals. The plates can be set at any distance, and all balls will be reduced to the same diameter.

The ends of heavy glass bottles can also be utilized to hold the sphere in the same manner as the iron pipe, but the glass bottle holds no advantage and is obviously somewhat hazardous.

The principle of the use of the iron pipe is not at all new, for many centuries ago the Chinese artisan used circular sections of bamboo for sphere cutting. A section of bamboo would be split lengthwise to form a half circle trough. The trough would be charged with some abrasive like sand, garnet, or impure emery powder, and the ball worked back and forth by hand.

The final shaping was carried out on the end of a complete circular section of bamboo. Since bamboo is readily available in various size sections, the method proved effective, except that the bamboo is not as uniform in diameter as modern iron pipe.

The most popular sphere machines are the double spindle, double motor machines such as this unit built by Pomona Lapidary.

Polishing Spheres

The method of grinding spheres between two bare iron pipes is well known and in common use. Some operators cover the ends of the iron pipe with heavy felt for the polishing operation. This method is quite satisfactory and is in wide use.

However, some sphere cutters prefer to carry out the final polishing operation using hollow wooden tubes. This goes back to the original technic long used in the Orient for centuries past. Bamboo was and is still used in the Orient in sphere work.

The hollowed wooden tubes need not be covered with felt and they have the advantage in that they are fairly soft and allow the polishing agent to work into the porous wood. A separate set of wooden pipes should be reserved for each polishing agent. Air floated tripoli is widely used; it is satisfactory for practically every gem material, and is low in cost.

Almost any variety of wood may be used. For some operations some sphere cutters prefer hardwoods like oak, but in general fir and pine are the favorites. The polishing technic is the same as when iron pipes are used. One tube is revolved by power while the second tube is held in hand. Small spheres, not over 2 and 3 inches in diameter, may be readily ground and polished on horizontal running tubes. The larger and much heavier tubes are best worked on vertical running pipe.

George Mathieu has made spheres for years and it is problematical whether his greatest achievement is this huge, double-spindle sphere machine or the massive sphere Mrs. Mathieu is cutting with it.

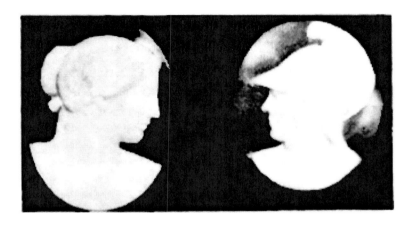

Shell cameos carved by Joseph Morgan.

Cameos and Cameo Cutting

E. F. MONTGOMERY

Cameo is a subject about which comparatively little has been written, and much of this deals with the cameos of years past. Writings on the more modern type of cameo, which is usually shell cameo, is limited to scattered and often contradictory papers. Practically nothing has been written in modern times regarding the technic of cutting shell cameo. Most of the modern papers regarding the subject consist of praise for the usually unknown artisans for the fine workmanship, and a description of various museum collections of cameos.

Derivation of the word cameo is not certain and may be defined as a small base relief or raised finger, usually executed on some shell or mineral substance. Cameos should not be confused with intaglios which are sunken below the surrounding surface.

First Attempts

After spending some futile time in an effort to view some well-known cameo collections, and to meet with someone who knew the technic of cameo cutting, I decided to launch forth on my own efforts. I dug out the smaller of my old wood carving tools, and used the comparatively solid mineral howite in my first serious attempts at cameo carving. The engraved figures appearing on our postage stamps proved excellent as a guide for copying.

Finally when the time came when I felt I wanted to start on shells, considerable difficulty was experienced in locating a source of supply of suitable rough material. After a search I found excellent rough shells a few miles from my home.

Shells Used

However, the best type of shell for cameo cutting is obtained from around the islands in the Indian Ocean, and various other regions in the South Pacific war zone. Probably the most treasured of all shells used in this work is the Casis Rufa Linn, better known as the "red helmet" or "bulls mouth." Even prior to the war very little high-grade shell cameo was imported, due to the lack of demand. Most of the shells imported were for collection and study purposes. The finer and larger shells were generally sent to the cameo-cutting centers in Europe. The "red helmet" has a soft creamy outer layer over a reddish orange background; it does not fade or "part" readily.

Next in importance for cutting are the Casis Madagacarencis and the Casis Cameo Stimps. These are commonly known under various names, including "cameo helmet," "black helmet," and "queen helmet." This type of shell is also found along the east coast of North America from North Carolina to the West Indies. It has a snowy white outer layer with an inner layer of various shades of dark brown, making excellent contrasting colors. Deeper down lighter shades may appear, thus offering added possibilities in executing various designs.

The Casis Tuberosa Linn ("sardonyx helmet") presents a porcelain white outer layer over a light brown or buff inner layer. When the outer layer is carved away, the inner layer takes on a burnt orange color of pleasing appearance. I am well pleased with this type of material. The shells in this group which I have worked have given a little trouble due to "parting," but I feel that this is due to poor quality material rather than an inherent defect in the shell.

The Craft of Jewellery Making

Raymond Addison, San Jose, California, carves "photographic" cameos by carefully controlling the thickness of the white layer so that more or less of the dark background shows through, creating a halftone effect.

The Casis Cornuta Linn ("horned helmet" or "yellow helmet") has a white outer layer. The shell makes a very beautiful cameo, but has a tendency to parting. It is found in the waters around Japan and the Philippines and the West Indies. It is said to occur with a pink inner layer, covered by white, which would make most excellent carving possibilities.

The Strombus Gigas ("fountain shell" or "queen conch") has a thick outer layer and an inner layer of beautiful pink. Its thickness lends itself to much higher reliefs than the helmets. The former may be reversed to present either a white or pink background.

There is no definite abrupt division between the two colors as seen in the helmet type. The pink and the white tend to blend in together. The Strombus is found in the waters off Florida and the West Indies, and is the largest snail found in American waters. This shell, while finely colored, and probably the least fragile, does have the defect of tending to fade, and must not be exposed to strong sunlight.

Cutting Methods

The methods used in carving cameo are similar to the technic once used by the writer in wood carving. First the shell is cut into small oval, round, or square blanks. While shell can be cut with an ordinary hack saw, I soon learned that my regular diamond lapidary saw was much more effective. The "resaw" is effective for working out the various shapes desired. An effort is made to obtain as large a blank as possible as trimming is generally indicated to eliminate thin or uneven spots.

The desired shaping blank is then attached to a small, flat board about 1×8×8 inches. Ordinary sealing wax may be used for this purpose, but care should be exercised to avoid overheating of the shell as this may cause future parting.

A general outline of the design is sketched or traced on the surface of the shell. A small hand grinder is then used to remove the white outer layer of the shell, outside of the design portion, to a point where the inner layer begins to show. Next the small wood carving tools or any engraving tools found suitable, are used to work down portions of the outer layer in the design. This is the most tedious and painstaking of the work. As many as 50 hours or more may be spent in executing a fine design or figure. The small gouges or wood-carving tools are of high-grade carbon steel, ground to various shapes to suit the needs. The writer uses some 20 or more of these fine tools, the cutting edge of which is only about one-sixty-fourth of an inch long. Another useful tool has a U-shaped cutting edge, and is less than one-thirty-second of an inch across the width of the U. It is obviously important to keep these tools sharp, and an optical loup or low-power magnifying glass will be found helpful in sharpening the points. The tools should be attached to a thick handle to enable applying heavy and steady pressure.

Four hand-carving tools and the carving board used by E. F. Montgomery for working cameos. Rough shell is shown at the left. Cameo carving requires skill and patience but is most rewarding.

While most of my cameo work is done with the naked eye, a jeweler's loup or glass will be found helpful in some types of work. Much of this is, of course, dependent upon the eyesight of the operator. In working upon a delicate face design as much as one-thousandths of an inch may change the entire expression of the face.

Hand Grinder

It was suggested by Dr. Dake that I try the type of hand drill and mounted points used by dentists. This technic was found better than the large and heavy hand grinders on the market, but hand work must be largely depended upon to give best results. The power tools are hard to control, and the tendency is to cut too deeply and perhaps ruin a nearly finished piece. A limited amount of cutting and carving can be done with power points, but the human hand must guide the greater part of the work.

Sanding

After the tedious carving has been completed the cameo is sanded lightly on the back, and perhaps slightly on the front to remove any tool marks on the raised figures. The sanding is done by hand using 400 grit (or finer) silicon carbide cloth. A flexible steel ruler is handy, the abrasive paper is wrapped around the strip of flexible steel. Polishing is done in the same manner, using a piece of felt cloth soaked in a paste of tin oxide and water. The carving itself can be polished, using a toothpick or similar point with a paste of tin oxide.

Machine Made

Many of the inexpensive cameos offered on the market today are either machine cut or made of some pressed material like glass or plastic. The pressed types are often of "doublets" or "triplets," cemented or pressed together, thus giving different contrasting colored materials. However, these machine-made articles can be readily separated from a well cut or even poorly hand-cut cameo. Prior to the war the pressed class cameos often sold for as little as ten cents each.

Possibilities

Cameo cutting is not as difficult as this paper may make it seem. Only a small amount of equipment is needed. Many of the steel points can be ground to shape and size using odd tools. The lapidary grinding wheels are suitable for this purpose. With a little persistence and patience the average gem cutter with good eyesight can do creditable work. Do not expect your first cameo to present perfection; that will come with time and experience.

Handy Lapidary Tool

For various special work and jobs, the gem cutter will find the flexible shaft hand grinding tool of considerable utility and value. The flexible shaft we refer to is the type having a pencil size hand-piece, similar to those in wide use by dentists, jewelers, and engravers. It is quite obvious that for fine work a pencil thickness hand-piece is indicated. The thick and bulky hand grinders are all right for certain jobs, but not so well suited for close and delicate work. It is like trying to write a fine and neat script using a pencil or pen one inch or more in thickness. Try it once.

These flexible shaft hand-pieces are readily available from supply firms and at reasonable prices. Two styles are available. In one the flexible shaft is attached to the armature shaft of an electric motor. If you use this type of shaft better results will be had for most work if a motor is used having a speed in the range from 3,400 to 3,600. The normal speed electric motor, as used in the lapidary shop, is generally around 1,750. To operate the flexible shaft, the motor need not be high-powered, one-fifth or one-sixth horsepower or less will suffice.

Flexible shafts of this type, complete with motor and hand or foot-operated switch, are also available, and these range in price from about $16.00 to around $25.00 or more, depending on equipment and accessories.

Most of these flexible shaft operated hand-pieces will take small size mandrels and a great host of points of this kind are readily available. These include hundreds of sizes and shapes of silicon carbide points and wheels, diamond points and wheels in many sizes, steel drills (called burrs by the dentist), felt polishing buffs, etc.

Equipment and tools of this kind will be found invaluable in executing fine jewelry work. One can hardly expect to do fine cameo carving without the aid of this tool. With the diamond points, holes may be drilled in various hard gems, provided the material is not too thick. Or in the case of thick work, the hole may be worked from both sides. Diamond points of this kind do not cut clearance, as in the case of a regular diamond drill, hence there is a limit to depth of hole. Soft gems, like turquoise, calcite, malachite, and others may be quickly and easily drilled with the steel drills, the round shapes. The steel drills, or burrs, are available in dozens of sizes and styles.

The hand-piece will also prove invaluable to those who wish to carve objects from various gem materials. The points for use in the hand-pieces referred to here are available from lapidary supply firms, dental supply houses, and jewelry supply firms. The grinding wheel or buff size that may be used in this tool is limited to about 1 1/2-inch.

When the diamond points and wheels were first introduced some 10 or 12 years ago sales were limited and prices ranged from about $5.00 to $12.00 each, making them a costly item. Now with mass production, prices have been greatly reduced. Used with care, as they should be, each point or wheel will give a remarkably large amount of service.

Tumbled gems can be made into colorful jewelry. The baroques are cemented onto bell caps which can be attached to chains and other findings to make bracelets, pendants, earrings, and other pieces. No soldering is required and the only equipment needed is a pair of long nose pliers and some epoxy cement.

The Tumbled Gems

During the past few years the method of finishing gem stones by the use of a barrel and abrasive has attracted wide interest. This is not at all new in the lapidary industry. A similar type of equipment was in use about 100 years ago, when they were referred to as "rumblers," no doubt in reference to the noise made.

Here are the three basic types of tumblers: Top, a multiple chamber, fixed drum tumbler. Pomona Lapidary. Above, an open-end tumbler. S. E. Landon Co. Right, a removable drum tumbler. Don Bobo Lapidary.

Many seem to be under the impression that the operation of a tumbling barrel is a simple matter, and that finely finished stones can be produced quickly. This is not exactly true. While it is correct that anyone without any knowledge of gem stone cutting can feed the rough materials and grits into the container, turn on the motor switch and start operations.

Attention may be called here to various requirements in order to operate this type of equipment in a satisfactory manner.

Many have predicted that the tumblers are merely a passing fad, and will soon die out. There appears to be at least some logic to this statement. It is obvious that for a collection of fine and valuable gems, the tumbled item hardly has a place here. No one would dream of tossing a handful of valuable rough opals, turquoise, agate, and similar items into the maw of the tumbler.

As a general rule only mediocre or the poorer grades of gem materials are trusted to the pounding and rough handling in the tumbler. Hence it will be seen that only a very limited number, if any at all, of tumbled gems would have a place in a collector's cabinet of fine gems. Only the skilled human hand can properly finish a valuable piece of rough gem material to the best advantage.

The Craft of Jewellery Making

Many designs are used for gem tumblers. These large machines will handle many pounds of material at one time. Top, a fixed drum tumbler the shaft of which is used to drive a wooden barrel used as an open-end tumbler. Wood barrels are preferred by some, particularly for polishing. Bottom, a large open-end tumbler with approximately 30-gallon capacity. These machines were made by Dan Powell, Portland, Oregon.

There are certain mechanical requirements that must be met with in order that a tumbler will function. This means that the barrel must be of a certain minimum size, and must be fed a certain minimum bulk and weight of fragmentary gem materials, grits and water, or whatever technic may be used. This is to enable the contents to "roll" and slide around, with the resultant friction doing the abrasive work. In short, some seem to get the idea that a satisfactory tumbler may be made from a tin can plus a handful of gem grits and abrasive. They just do not work this way. Try it yourself. This is why all specifications given for this machine involves a container having approximately a 5 gallon or larger volume. This means an abrasive charge of several pounds of grits to handle say a 20 or 30 pound charge of gem fragments.

Here are some of the objections, disadvantages and advantages that have been voiced for this tool. It most certainly is an easy means of placing a fairly good polish on a large number of pieces. Some find the tumbler handy for removing the matrix materials from small sawed slabs and the like, and then complete the work in the usual manner on the grinding wheels and polishing buffs.

On a commercial production scale, small diameter sawed slabs may be placed in the tumbler, with polishing powder and water, and given a reasonably good polish. Small agate slabs are often finished in this manner and this enables the supply house to sell them at a much lower cost since we have here a labor saving device, eliminating dopping and hand polishing.

Some have objected to the "messy" part of the work. In changing from coarse to fine grits, and then to the final polishing powder, it is necessary to clean the grits from the material as well as the barrel itself. This is not always easily done. The supply houses that do this work on a commercial and custom basis usually operate a whole battery of tumblers, a dozen or more may be in operation at one time. In this case the barrel itself need not be cleaned since the same abrasive grit is used in the same container for the next batch to run. It is not feasible for the home gem cutter to use this technic. Attention may be called to the fact that plumbing can be plugged when heavy silicon carbide grits are dumped into a household drain. This is why the tumbler has been dubbed the "plumber's friend" by some users.

From a commercial standpoint, these unique gems are widely used in costume jewelry, including items like bracelets and necklaces, where a dozen or more may be mounted into a single piece. Special caps are sold by the jewelry supply houses to mount the stones by merely cementing the metal caps to the stone. At the present time rather large quantities of tumbled gems are being used for this purpose in the costume jewelry manufacturing trades.

The number of hours that a tumbler must run in each step of the work will vary depending on various mechanical factors, like type of materials being run, size, speeds, grits, and the like. It is customary to stop the barrel at intervals of some hours and examine the progress of the work. But in each case a good many hours are required in each step. This is not a one day or overnight job.

Types of Equipment

A number of different types of tumbling barrels have been described. The most popular, and what appears to be the most efficient is the hexagonal shaped metal barrel. These may be anywhere from 12 to 16 inches in diameter, with a same length or about twice the diameter. In some commercial built units the barrel may be divided into two or more sealed compartments to enable two or all three of the operations to be carried on at one time. For the home worker this seems to be quite an advantage. The metal barrel must have an opening on either the side or the end to enable charging, removal, and cleaning. Complete units of this kind, ready to use, are being currently offered at from about $45.00 to over $100.00, depending on size, equipment and the like.

Some homemade units are simply a suitable size metal drum, an oil container for example. A feeding hole is cut at one end, and the drum mounted so it may revolve, resting in a horizontal position. Wooden barrels may also be used but metal is preferable due to more certainty in cleaning out abrasive grits.

Since most of the abrasion is done not by the falling of the material but by its sliding due consideration must be given to the speed at which the barrel is turned. In many cases failure to get satisfactory results is due to improper speed at which tumbler is revolved. One manufacturer of a 30-inch barrel advised a speed of about 20 r.p.m. with the barrel half filled with total charge of gem material, grits, and water. In all cases the unit is operated at quite low speeds.

There are many mechanical factors involved that will determine the best speed of efficiency, and this will include size of unit, type of material being worked and the amount, bulk of total charge, amount of water, and the like. Most tumbler operators determine the best speed by experimenting, the progress of the work being observed from time to time, and from this the best speeds for the different operations can be learned through experience.

There are two general ways of mounting the barrel. Since

the container and the charge is heavy, many have found it advisable to fasten metal flanges on the ends of the barrel to allow it to rest on metal rollers or bearings; a removable barrel being handy for cleaning and charging. Or flanges with a short axle and pulley space may be bolted or welded to the two ends of the barrel. Since the unit operates at slow speeds a comparatively low powered motor may be used. To dampen the noise, a metal barrel may be lined with some material like sheets of sponge rubber cemented into position. Most commercial motors operate at the standard speed of 1,725 r.p.m., so this speed must be reduced by a proper pulley system at the motor shaft and at the barrel, or a system of speed reduction between the two. Slightly lower speeds are generally advised for the polishing.

A small, table top, hobby tumbler with a removable barrel. This simple design is made by many manufacturers and because the barrel lifts off, is easy to load and clean. The hexagonal barrel is considered the most efficient though other shapes are used widely. This machine is distributed by R & B Art-Craft Co., Los Angeles, Calif.

Abrasive Charge

A good deal has been printed about the best abrasive charge to use, and advice on this seems to vary with every operator.

It is generally advised that on the average the barrel should be about half filled with the final charge, including the water. In some cases this is increased slightly or reduced, but it is to be remembered that the unit will not function unless the proper "sliding" action takes place. This means that if the total charge be too small or too large the required "rubbing" action will not take place. Hence where the charge is too thick and heavy, like lack of water, the charge will merely roll around and not abrade or polish properly.

Various advice has been given on the type of materials that may be worked. While certain materials may be worked at the same time, in general it is advisable to work at one time materials having a similar hardness. Where a hard gem like agate is worked with soft ones like malachite, the latter material is likely to be completely or badly ground away.

The Craft of Jewellery Making

Earrings made from tumbled gems. All that is needed are two tumbled stones, two bell caps, two jump rings, and a pair of ear wires. The caps are cemented to the stones with epoxy cement, then fastened to the ear wires with jump rings, and the set is complete. This is the simplest form of jewelry making.

Various methods have been given as to the abrasives used. Silicon carbide grits are universally used. The general use seems to be of 100 grit for the first rough grinding. This is followed by a run with 320 grit. There can be many variations here. One can start with a much coarser grit and work down to 600 or even finer grit. But two grit runs seem to suffice for all ordinary work.

Like the technic advised in cabochon work, it will be noted that the polishing operation will proceed faster and more effectively when the final grinding (or sanding) operation has removed all deep scratches. This theory applies to tumbling. Many operators note that where the final abrasive tumbling operation is carried out with a 600 grit, the final polishing will take place faster, and the work come out with a better polish. In any case it is obvious that where the material placed in the container with a polishing agent still presents deep scratches a long time will be consumed in polishing; it can be done, but the lesson learned in cabochon work may well be heeded here.

There is a definite relation to the grit charge and the type and weight of the material to be worked. In general about 5 pounds of silicon carbide grits are used with 20 to 25 pounds of rough gem fragments or sawed slabs. Some advise more and some less.

There has been a good deal said about the advantages and disadvantages of using items like steel balls in the grits. Commercial balls made of silicon carbide have also been used. Items like sands, sawdust, etc., also have their advocates. Iron balls are in wide use in what are called "ball mills." This is a commercial process where various materials, like ores, are ground fine before being run through the recovery mill. Grinding mills using steel rods, called rod mills, are also widely used for grinding. In the rod and ball mills grinding the material to powder is the sole objective. This phase is a matter of debate, and the reader may experiment if desired.

Final Polishing

It is in the final polishing operation that the most difficulties are encountered. Almost any polishing agent may be used in the final polish. It must be kept in mind that unless the grinding grits are completely washed from the container and the gem materials, it will be futile to expect a satisfactory final polish. It would be like attempting to attain a perfect final polish on a felt buff that is contaminated with silicon carbide grits. Where the gem materials being tumbled present pits, fractures, cavities, and the like, it is an easy matter to carry abrasive grits over into polishing operation.

Practically every tumbler operator has his own favorite polishing agent, and these include all the polishing agents that have been used in cabochon work. In the final analysis it would appear that the inexpensive air floated tripoli is about as good as any. Tripoli has been an almost universal polishing agent in the lapidary industry for centuries, and long used before the more modern polishing agents were invented. Tripoli is still good, particularly on modern air-floated product. The polishing operation will take longer, as a rule, than the grinding operations. Like in cabochon polishing, all things being equal, the tumbling polishing operation will also be dependent on the scratches to be removed, the depth of the scratches and the like. But in any event polishing will take longer, perhaps twice as long on the average.

Many operators advise that some detergent cleaning agent be added to the polishing charge. This is to promote better contact between the polishing powder and the gem material. Many of the commercial detergents have been advocated. Some favor tossing in a good sized piece of soap. Some operators advise the use of as many as three and four entirely different polishing operations, following from two to four grinding operations. This seems tedious, but it is no doubt effective. One would perhaps be obliged to start operations in mid-summer to get baroques in time for Christmas.

Most speeds advised for polishing are about half the grinding speeds. In short, the highest speed is generally for the first rough grinding, and as the different "finer" steps are reached the speed is slowed, ranging from about 25 to 30 r.p.m. down to 10 to 15 r.p.m. in the final polishing work.

Above, a small, table-top, open-end hobby tumbler with a capacity of about a quart of gem material and slurry. S. E. Landon Co. Left, an old automobile tire used as a tumbling barrel. The tire simply rests on the two driven shafts which supply the drive and the speed reduction.

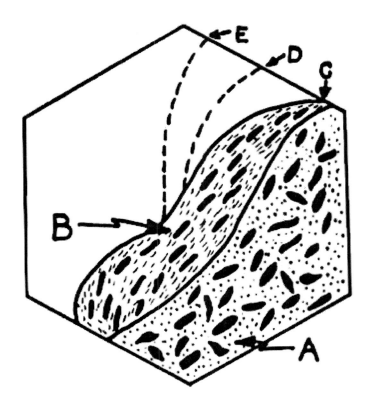

THEORY OF THUMBLING

A is nearly stationary material. B is the sliding layer where actual grinding or polishing takes place. C is the point where sliding starts at about 15 r.p.m. D is the start of sliding at 25 r.p.m. E is the start at 35 r.p.m. Speed is usually set so that sliding starts at about C for polishing and between C and D for grinding. If the speed carries the material to E, it will drop instead of slide and the clicking sound will indicate the stones are hitting together instead of rubbing. The speed should be adjusted for all operations so that maximum sliding action is achieved for grinding. A milder action is desired for polishing, and hence a slower speed.

Attention is again called to the fact that if any of the abrasive grits are carried over into the final polishing operation, the work will continue to show only a fine "satin" finish no matter how long the tumbler is operated. The polishing charge must be free of coarse abrasive grits in order to bring a proper glossy finish in the gems. Or if the polishing operation has you licked, it is still possible to take the nearly finished baroques to the polishing buffs, a few applications of hand effort and the job will be done.

Some experimental work has been done with the idea in mind of making use of a container that can handle a smaller charge in an effective manner. These containers include the use of a long but small diameter barrel, with the barrel operating on off-center shafts. This is to allow the charge to slide from one end of the barrel to the other, as it is slowly revolved. A proper mechanical arrangement of this kind should prove effective.

Some advise the use of "additives" to the grinding charge, and these include materials like the sands used in commercial sand blasting. In the place of detergents, some prefer the use of common bicarbonate of soda, using a few pounds to the charge. It would appear that all of these agents would be more or less effective in both grinding and polishing operations. They would also aid in the prevention of slimy sludges which tend to slow down abrasive polishing action.

THEORY OF TUMBLING

There is a wide open field here for experimenting unlimitedly for those who may be so inclined.

Some of the supply houses producing baroques on a mass production scale offer custom work. The fee for finishing rough fragments is quite nominal compared to the standard charges made for custom cabochon work. Some commercial establishments operate quite large tumblers, some of them requiring a charge of hundreds of pounds of gem fragments. Many find it most satisfactory to send their fragmentary materials, not suitable for cabochon cutting, to these commercial tumblers. Small sawed slabs may also be handled in this manner. Large diameter slabs would likely be broken when tumbled, except in the case of very tough materials like jade.

Tumblers, or "mills" are they are termed, are widely used in various industries, including the mining and foundry trades. The smaller size iron castings are usually finished in tumblers, and here the operation is usually done dry, with abrasive sands, iron balls, rods, and the like. Here again the work is done, not by the material falling, but by sliding as the barrel revolves.

For large scale commercial tumbling of gem materials, some operators have made use of the equipment commonly used in the foundry casting industries. These rugged and versatile units can be readily adapted and utilized to good advantage for gem tumbling work.

A typical, vertical shaft, flat lap used for working large flat pieces such as display slabs, bookends, transparencies, etc. The lap is usually cast iron but can be replaced with a variety of lap surfaces for grinding, sanding, and polishing. Photo, Covington Lapidary Engineering Corp.

A large over-arm flat lapping machine used by the National Museum, Washington, D.C., to prepare display slabs. Grinding laps, sanders, and polishing buffs can be attached to the vertical spindle which swings over the work.

The Horizontal Laps

The horizontal metal laps illustrated here are inexpensive and useful, especially for finishing large flat surfaces on specimens. There are a number of excellent ready-made units of this type. These are sold at a reasonable cost.

The horizontal running lap is inexpensive to operate, as only cutting grits and polishing powders are needed. The cutting and polishing of a large flat surface need not be limited to minerals of the gem class; often a compact specimen of a metallic sulphide can be greatly improved by polishing one of the flat surfaces. Often the polished surface will bring out interesting surface features not otherwise readily noted.

For facet cutting the horizontal laps are generally run at low speeds—around 100 r.p.m.—for both cutting and polishing (eight to ten-inch lap). However, for roughing down a large specimen high speeds will be found more effective and quicker. The higher speeds are also more effective in the final polishing operations.

Most horizontal running units may be made with interchange pans and metal laps, so when passing from one grit to another it will not be necessary to clean the accessories. Simply wash the specimen free of grit and change the iron lap and the pan carrying the abrasive mixtures. A separate lap is kept for polishing—generally a metal lap with a heavy layer of felt cemented on the surface will suffice for polishing the flat surface, or a wooden lap coated with felt can be used.

A well designed over-arm lapping machine built for the amateur. The spindle swings back and forth over the stationary work. This machine is particularly useful for large slabs, table tops, etc.

Three iron laps and separate pans are very convenient. For roughing down a large specimen to a flat surface, No. 100 grit (or coarser) can be used. This is followed with the iron lap and pan carrying No. 200 and then to No. 400 grit and finally the polishing. The silicon carbide grits are mixed with water and allowed to remain in their pans and used repeatedly.

Some commercial units are fitted with ball or roller bearings throughout, which give better service and less noise in operation. The laps used in the grinding operations are generally made of ordinary cast iron ("grey" iron castings) and are not over one-half inch in thickness. They should be fitted to the vertical shaft by means of a tapered fit. This leaves the lap surface free of any projections and enables the easy change of laps without the use of tools. If the tapered fit is reasonably accurate the weight of the lap itself will be sufficient to eliminate any tendency to slip on the shaft.

Many home lapidarists operate units of this type with considerable satisfaction. A lap having a 12-inch diameter will enable grinding and finishing a flat surface of large size, as without any projection on the lap surface the specimen can be worked across the entire face. Generally, speeds of around 300 r.p.m. are used, but this tends tb throw the abrasive off the lap surface. The abrasive can be kept piled in front of the specimen by feeding, by hand or by mechanical aids. Higher speeds cut faster, provided some provision is made to keep abrasive passing under the work. If the lap surface runs dry the specimen may suddenly "stick" and be jerked out of the hand.

The possibilities of the horizontal running lap are many and varied and additional uses will suggest themselves to the experienced operator. Much can be said in favor of this type of lapidary unit. For cabochon cutting it is nearly useless.

A vibrating flat lap built by Fulmer's.

Cutting and Polishing Transparencies

Thin sections of transparent or translucent gem minerals can often be conveniently displayed as "transparencies" by using daylight or artificial light for illumination.

A number of methods will suggest themselves, a few will be detailed here. Large slabs of agate can be very effectively displayed in this manner. The sections need not be cut very thin, merely sufficiently thin to permit the passage of light through the specimen. This will be dependent largely upon the material itself and the intensity of the illumination. Some specimens show up well if cut ¼-inch in thickness, others must be sawed or lapped down further.

A rough irregular shaped specimen can be displayed with the edges in the rough form and not detract from the beauty of the pattern. Generally the diamond saw is used to saw the section to proper thickness. The horizontal lap or the side of the grinding wheel will serve to further reduce the thickness and remove any deep scratches left by the saw. Since the specimens are often mounted between glass, a high polish is not essential. Or as an alternative to polishing, the flat surfaces can be given a light sanding and then given a coating (both sides) of Dake Varnish. Since the transparencies mounted behind or between glass are not handleed, the varnish method gives good results and only very close observation will indicate the specimens have not been polished. Moreover, polishing a large flat section on both sides entails considerable labor; the varnish will eliminate much of this labor.

The Vi-Bro-Lap. These machines work flat surfaces automatically. The inertia of the heavy stone holds it relatively still while the lap vibrates underneath. A variety of grits, polishing media, and lap surfaces can be used.

Work is usually held by hand on a flat lap and moved around to distribute the grit and bring fresh grit under the stone.

Picture Frame Method

One popular method of displaying small thin agates, and similar gems, is mounting in a frame used for photographs. Various size frames can be utilized according to the fancy of the individual.

The specimens are placed on a piece of light or black colored cardboard, the same size as the glass in the frame. A mark is then made around the outline of the specimen, and holes are cut to correspond. The specimens are then placed behind. The cut stones or thin specimens are now mounted between two pieces of glas. The light passes only through the specimens, thus making them stand out in contrast. The frame can be viewed by hanging in the window or held toward any light. Small size frames of this kind make admirable desk ornaments.

Illuminated Frames

The above method can be used in the same manner except that boxed-in illumination is used as a source of light. Discarded commercial illuminating advertising signs can be effectively utilized. Generally signs of this kind are made of metal, portable, wired for electric globes, and fitted with a glass on the front. The advertising sign on the front can be removed by scraping the paint from the glass. A similar size section of glass is cut to fit the back of the cardboard same as is used in the picture frame arrangement. This method offers numerous attractive and artistic possibilities. A frame of this kind measuring 12×24 inches can be filled with a very colorful array of cut stones and irregularly shaped specimens. At least an inch of dark space should be left around each specimen; packing in too many is not effective. Illuminated frames of this kind could be used to good advantage in commercial display windows at night; intermittent illumination attracts attention.

A slanting shelf of ground glass can also be used to display transparencies. Special lights are placed behind the ground glass to give a pleasing diffused light. Cardboard is not needed to eliminate any excess light, the thin specimens are placed directly on the slanting glass shelves.

Mounted Transparencies

William B. Pitts, noted lapidarist of San Francisco, California, has suggested a means of mounting transparencies which holds wide possibilities. By this method a number of semi-precious gem materials can be arranged to be viewed as transparencies. The thin section specimens can also be cemented on lantern slides and thrown upon the screen.

Chiastolite crystals are next to impossible to polish in a satisfactory manner. Mr. Pitts, in experimenting with this material, employs the following technic of mounting.

The Craft of Jewellery Making

Obviously the same method can be used for a number of similar materials.

The delicate patterns seen on cross sections of chiastolite are due to inclusions of carbon, and no two are exactly alike. The mounted frames of glass can be viewed by holding toward any light, projected on the screen, or used in an illuminated frame. The low-priced projectors fitted with *Polaroid* can be used to project various thin sections on the screen under ordinary light or polarized light. Some gem minerals give remarkably beautiful interference colors when projected with polarized light.

Some flat laps are fitted with a lever so that large heavy pieces can be swung back and forth easily. Sometimes the stone is fastened to the lap and the grinding tools mounted above, under the movable arm.

Thin sections such as these transparent agate slabs can be mounted and illuminated from the back for startling effects.

Sawing Crystals

The crystals of chiastolite are first cut across the face by the diamond saw to a thickness of about 1 mm. (1 mm. = 1/25-inch) or thicker. The sections are then ground by hand by holding on the side of the grinding wheels or the horizontal lap until light will pass through the section. The thickness of the section will be dependent upon the opacity of the material. From one to three mm. will be the average. The specimen is then made smooth on both sides by lapping on the iron lap with No. 500 or No. 600 silicon carbide grit. No attempt is made to polish as this is not needed. If a horizontal lap is not available, the section can be worked down by hand labor using a plate-glass slab and a mixture of water and grit. Hand labor is not tedious unless the sections are very thick to begin with.

Mounting on Glass

The thin sections of chiastolite (or other material) are then cemented to a suitable size section of ordinary window glass. Ordinary varnish can be used, but Canada balsam will be found much more effective. The Canada balsam is the liquid kind, the same as is used in mounting thin microscopial sections of rock or minerals. A drop of Canada balsam is placed under each small section and the excess pressed out. The glass slab with the arranged and cemented specimens is then "cooked" to dry the cement. A temperature of nearly boiling is necessary to properly dry the balsam. A low temperature hotplate or oven will suffice. Or the plate can be left under a strong electric light and the heat will dry the balsam in a few hours. If the balsam is not cooked enough the specimens will tend to slide about on the glass; if this happens, further heating is indicated.

Grinding and Backing

If some of the specimens appear too thick further grinding can be done on the cemented specimens. The entire face of the slab can be ground on the horizontal lap. If the edges of the glass mount become "frosted" from the abrasive no harm will be done. Grind until light passes through the thick specimens.

An opaque backing of some kind must now be placed between the thin sections of chiastolite or other material. William Pitts advises the use of a thin layer of *Handee-Wood*, a plastic similar to the familiar plastic wood. A thin layer of this substance is spread over the specimens and the excess squeezed out by pressing down on a glass plate. Any surplus material covering the face of a specimen can be scraped off with a knife. About two hours will be needed for the setting of the *Handee-Wood*, when the edges and sides ocan be trimmed with a knife.

Transparencies of agate cut and mounted to make a picture of potted flowers. Photographed by rear illumination. About 4×6 inches. Made by W. A. Burt, Portland, Ore.

The entire surface of the backing and specimens can be varnished over with ordinary or Dake Varnish. To better protect the specimens a glass plate can be cemented over the backing, using liquid Canada balsam, and again "cooking" the plate.

W. Nelson Whittemore, well known lapidarist of Santa Barbara, California, calls attention to the fact that sealing wax is brittle and under some conditions may tend to peel off from the glass. Moreover, sealing wax is somewhat expensive to use for this purpose.

Whittemore suggests that a mixture of equal parts "solid" black tar and rosin be used as a substitute for the sealing wax. The tar and rosin are mixed by heating and can be cast into rods or sticks and used in the same manner as sealing wax. The tar and rosin mixture is much cheaper than sealing wax and is not as brittle. The tarrosin formula can be varied to meet specific conditions; the more rosin added, the mixture becomes more brittle and has a higher melting point.

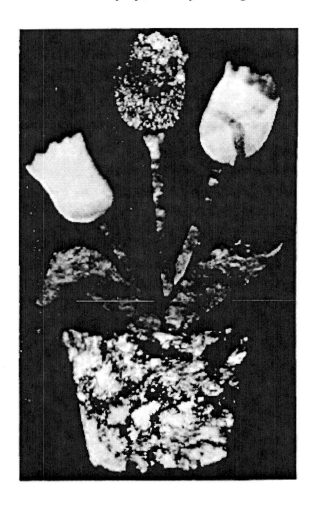

Tulips made from thin sections of colorful red moss agate, chrysocolla, and carnelian. The stones are mounted on glass (see text. Made by W. A. Burt, Portland, Oregon.

Since transparencies are often illuminated with electric

lights in metal cases, a backing mixture with a fairly high melting point is indicated, otherwise the backing may melt. Transparency cases are finding wide application in museums. Cranbrook Institute of Science, at Bloomfield Hills, Michigan, recently installed some magnificent transparency wall cases which are attracting considerable attention.

Transparencies—Burt

W. A. Burt of Portland, Oregon, has described a method of cutting and mounting transparencies in the form of various art objects. The Burt technic is described by him below.

In arranging cut sections to be viewed as transparencies, it occurred to the writer that there are many possibilities which have perhaps been overlooked by many gem cutters. Thin sections of gem minerals arranged into some specific pattern, as shown in the accompanying illustration, hold unlimited possibilities and will test the artistic ability of the cutter.

Cutting thin sections of gem minerals to complete designs as shown in the illustrations, is not a difficult matter. A little experience, care and patience will enable the home lapidarist to produce some really creditable work.

In the preparation of a flower or similar intricate patern, the writer first cuts each piece in outline in heavy cardboard. The outline in cardboard is then placed over the sawed section and with an aluminum pencil trace the shape upon the sawed gem material. The saw and grinding wheels are then used to cut to proper shape.

In cutting an intricate piece, a *Handee Grinder* and small silicon carbide grinding wheels will be found invaluable. However, it is possible to assemble patterns as shown here without the aid of any special lapidary equipment.

Various types of gem material will suggest itself for this purpose. Colors and pattern should all blend in together to make a harmonious and pleasing color scheme. Thin sections of chiastolite, mounted in glass frame, also make supurb transparencies.

The thickness to which the specimen should be sawed will depend upon the color and transparency of the material at hand. Material which tends to opacity should be cut thinner. In general the cut sections will vary in thickness from 2 to 6 millimeters. All mounted frames should be of uniform sizes, as this will enable mounting the frames in a large wall mounting, with ordinary electric lights behind the mount. The writer prefers frames 8×10 inches in size.

One of the advantages of transparencies is the elimination of tedious polishing of thin sections, since the transparencies are fixed to the glass with Canada balsam and varnished on the back with Canada balsam and xyol (equal parts of each, "Dake's Varnish"). However, the work will appear neater if the deep saw marks are removed from the surfaces of the sawed section. A light sanding will generally suffice to prepare the thin sections for mounting.

The cut and assembled sections are cemented to the glass with Canada balsam and baked in the oven at about 200 degrees F. for about five minutes and then allowed to cool. The glass with the cemented specimens is then mounted in a suitable wooden frame. The writer prefers to "black out" the space around the thin sections with ordinary black sealing wax.

Inexpensive wood frames to hold the glass may be obtained at Woolworth's where numerous standard sizes are available to suit the needs. The writer has over 30 frames of transparencies, similar to those illustrated here, in his collection of transparencies. One of the most complicated frames, a group of daisies in bloom, required over 50 separate pieces of gem material, and over 20 hours of labor.

The Craft of Jewellery Making

Butterflies made of transparencies of agate. This is a popular way to utilize matching pieces of stone. These were cut and mounted by W. A. Burt, Portland, Oregon.

The work shown here, while painstaking, is not at all difficult and can be executed by the home lapidarist. Designs of this kind can frequently utilize small fragments of gem slabs that may not be suitable for other purposes. Animals cut in outline (dinosaurs, etc.) lend themselves well to this technic.

Drilling Cabochons

The customary method of drilling holes in cabochons is by the use of various gauge hollow tubes, which are available from supply firms. Short lengths of the tubes are mounted in a small power-driven drill press, the work to be drilled is fastened securely with a clamp or held in place by some other mechanical means to leave the hands free to operate drill. The drilling abrasive can be held to the work by placing a small dam of putty on the drilling point.

The tube drill should be given a "set" so it may cut clearance in the hole, otherwise it will tend to bind and break. The "set" may be given by slightly tapping or cutting the end of the tube drill. A quicker and also satisfactory manner of cutting clearance is to slightly bend the end of the tube drill by holding a pencil against the drill when in operation. This slight "wobble" of the drill point will serve to cut clearance. In drilling it is customary to raise and lower the drill when in operation.

A special, high speed drill made especially for drilling stones. The arbor oscillates up and down automatically to pick up slurry. Used primarily with small tube drills. Photo, Covington Lapidary Engineering Corp.

Various types of automatic drill presses have been described in these pages. These are highly satisfactory for they automatically raise and lower the drill point, leaving the operator free to do other work. As a substitute for the hollow tube drill, a shingle nail or finishing nail of the proper gauge may be used with good results. These are study, inexpensive and readily available. A slight "wobble" is given the nail in the same manner as the tube drill.

The writer has watched operation of a clever home-made and automatic cabochon drill used by M. T. Green of Bend, Oregon. Green has drilled many hundreds of agates on his simple power driven equipment using a fine finishing nail as the drill, and 100 grit silicon carbide (mixed with light oil) as the abrasive.

Various operators use different abrasives, but 100 grit silicon carbide is highly satisfactory for drilling fairly large holes required in agate ornaments and jewelry. In drilling small diameter holes in more valuable minerals like opal, diamond grit is advised.

Soft gems with a low hardness like turqoise may be readily drilled with steel burs such as are used by dentists. These steel drills are available in various sizes and shapes, the round drill being most generally effective.

In recent years mounted diamond point drills for lapidary work have been developed. These are simply small angular fragments of diamond mounted securely on a steel drill shank. Various sizes are available from supply firms. Used properly in any standard gem drilling unit these mounted points are effective for hard materials, and will give long service. Being a delicate tool they will not stand rough handling, heavy pressure, or forcing. The directions given by the manufacturers should be followed carefully, the main object being to keep the drill point free of an accumulation of debris which interferes with the diamond point reaching the work.

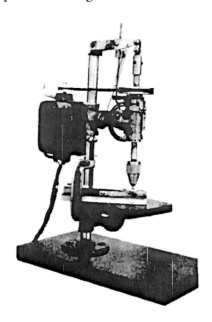

Any lightweight drill press can be used to drill stones. Without automatic equipment, the spindle must be oscillated up and down by hand. High spindle speeds are needed.

It is essential to use a high speed drilling machine capable of spindle speeds up to 5,000 r.p.m. There are a number of them on the market, and some are automatic or semiautomatic, raising and lowering the drill as it is in operation. Make sure that the diamond drill runs true in the chuck before placing in operation, otherwise the diamond or drill may break. The work should be clamped solid, and this is usually done by cementing the stone to a block of wood, using ordinary dopping wax.

Speeds of the various types of drills, and depending on thickness of hole, will vary from about 3,000 to 5,000 r.p.m. Start the hole by careful and gentle hand operation, raising and lowering the drill from the work, once the hole is started less caution will be required. As a lubricant for the drill and to aid in keeping the point free of debris a few drops of light oil are kept at the point of work. Some operators prefer a mixture of equal parts light oil and kerosene.

Bracelet Cutting

Bracelets for wrist wear, cut from various semi-precious gem minerals, can be worked by various lapidary methods. Perhaps the method use by the Chinese lapidarist is the most simple and effective. The Chinese often produce splendid bracelets from jade and nephrite. In selecting a gem material for bracelet cutting, the mineral should be tough, free of fractures, flaws, and cleavage. A gem mineral with a marked cleavage would be wholly unsuited to withstand wear as a bracelet, or even a finger ring.

Chinese Method

In working valuable jade, the Chinese have long since learned various methods of conserving the rough material. In hollowing out a vase, cup, or bowl, the Chinese artisan will first cut a small core to the proper depth. The core is sunk by the tube drilling and then broken out with a sharp blow from a hammer. The core can, of course, be used for cabochons or any other suitable ornament.

After the central core has been removed a larger size tube is used to take out a hollow cylinder of the desired diameter. In order to avoid breakage when taking out the cylinder, same should be undercut, working in the space left by the original core. The undercut can be made by the use of small and tin-mounted silicon carbide wheels on a hand grinder.

A block of rough gem material will also suffice for obtaining hollow cylinders of various sizes suitable for rings and bracelets. The same technic as described above can be applied, first removing a central core.

Tube Drilling

Tube drilling can be carried out by various methods, the power drill press being the most convenient and effective. The Chinese use ordinary iron pipe, various diameters, and as an abrasive, coarse silicon carbide grit serves as the cutting agent. Copper and brass tubing is also excellent. Diamond dust (bort) of about 120 grit is the fastest cutting agent for use in tube drilling.

In order to cut clearance the tube used for drilling should be slightly flanged on the working face. This can be done by gently tapping with a hammer. With large heavy tubes a chisel can be used to cut notches on the working face, and a flange bent down by hammering. The notches will also serve to better hold the abrasive where the cutting is being done.

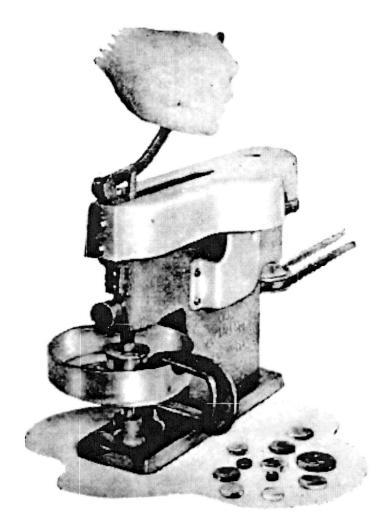

The Hillquist hole saw. This works in a manner similar to a gem drill but uses larger diameter tubes to cut buttons, rings, and similar pieces. The tools are metal pipes, usually copper or brass, that are charged wthi loose grits.

Plenty of lubricant should be applied at the working point. In drilling very small holes, light oil is mixed with the abrasive grit. For very large holes water is a satisfactory lubricant.

Grinding Blank

After a cross section has been cut from the cylinder it is necessary to round off the edges by grinding on the silicon carbide wheels. Holding the work by hand is unsatisfactory as it will be found difficult to keep the bracelet the same diameter throughout. If cut too deeply on one area it will be necessary to reduce the entire surface to this same diameter.

In grinding a bracelet (or finger ring) the Chinese does not attempt to hold the work against the grinding wheel. The work is revolved against the wheel and by this method it is not at all difficult to produce a symmetrical piece of work. A method whereby the grinding wheel revolves in one direction and the work in the opposite direction would be ideal.

The Craft of Jewellery Making

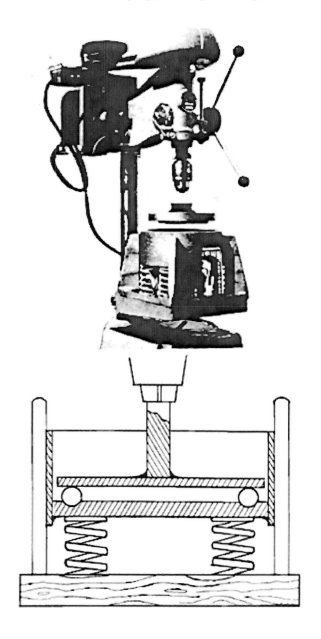

The bead mill is another special machine for making small spheres. The device can be used with any drill press. The one shown is the Crown.

Polishing

After the blank cut from hollow cylinder has been properly shaped by grinding, the deep scratches should be removed by the usual sanding operation. The inner surface of the bracelet or ring can be sanded by the use of various size felt cones, and small felt wheels using fine grit (220 or finer) silicon carbide or *Norbide*.

Final polishing is carried out on the regular felt buffs, holding the work by hand. The inner surfaces can be reached by small felt polishing wheels or cone-shaped polishing wheels. A cone-shaped polisher can be made by cementing felt or soft leather to a cone-shaped piece of wood.

Small felt buffs and cones are available from supply houses. These are intended to be used on small polishing motors and hand grinders. The regular polishing agents are indicated in this work.

Agate Rings

There are a number of gem minerals other than jade and nephrite which can be worked into bracelets and rings. Agate, free of flaws and fractures, can be utilized as this material lacks cleavage and is quite tough and tenacious. Some of the hard, tough, and compact varieties of massive garnet (grossularite) are well adapted for these ornaments. Other gem minerals will suggest themselves as being suitable.

In cutting a hollow cylinder for a bracelet due care and attention should be given to the final size to slip over the hand. The size of a finger ring can be enlarged by grinding on the inner surface, if there is enough material to permit same without breakage of ring. This can be done by the aid of small grinding wheels and buffs, mounted on a small mandrel, and help in a hand grinder.

Hand Grinders

Hand grinders are available from various lapidary supply houses and are available in several styles. One type is powered by a very small electric motor contained within the unit. Another style hand grinder has a flexible shaft which can be attached to the armature shaft of a small electric motor. Grinders of this kind find wide use in the home jewelry and lapidary shop.

USEFUL LAPIDARY NOTES

Cutting a Cabochon

All too often the amateur gem cutter is likely to fashion a gem in such a manner as to render mounting difficult or hazardous (to the stone). A gem cut in such a manner as to present thin and fragile edges brings up difficulties in mounting. In the case of cabochons, if the edges of the stone are lacking proper angle, it will be impossible to properly clamp the bezel.

A well mounted gem exhibits a mounting which is true at all points, with angles matching and lines joined in perfect unison. Obviously this cannot be accomplished if the stone is not cut true.

Preliminary to cutting a gem it is well to carefully examine the rough material at hand and decide the best angle at which to cut the first slab. It is customary, for a number of obvious reasons, to saw the first slab (the heel) slightly thicker than normal. Then each succeeding slab is cut according to indications.

Depending upon the nature of the material at hand the sawed slabs should be cut accordingly. Some materials will be cut thicker than others. In cases where there may be a good deal of paste, minute pockets and vugs present that will require future grinding, the slabs should be cut thicker than normal. By "normal" we mean on an average of about 3/16-inch thickness. As is well known, some gem materials like jade are exceedingly tough and large slabs cut only 1/16-inch will stand considerable punishment without danger of breaking. On the other hand, softer and more brittle materials like malachite and calcite onyx, and turquoise with matrix present, these should be cut thicker.

A little experience will soon enable the gem cutter to judge the best thickness for the material at hand, as no hard and fast rule can be set down. Tiger eye is another good example of a tough fibrous material that will stand lots of abuse and still not fracture. Materials that are needlessly cut too thick in sawing will require future added labor on the grinding wheels to reduce to proper thickness for cabochons. A slab of gem material, sawed too thin at the outset, may be rendered useless for cabochons. So to play it safe it is advised that the beginner saw fairly thick, even though this does involve added labor on the grinding wheels.

Next, having the slabs sawed: Check and mark out all good pattern areas to size and shape of the future cabochon. These areas and shapes may be outlined with an aluminum or bronze pencil—a piece of pointed aluminum or bronze wire makes an excellent pencil that will not wash off in the grinding operations.

Now we go to the handy trim saw where the outlined areas are cut off and the waste portions discarded. The roughed out blanks are then ready for the grinding wheel operation. Now begins the important task of grinding to proper shape, the removal of undesirable areas, and reduction of thickness if the slab has been cut thick. Next you must know how to please the fellow who is to place the stone in a mounting. This includes placing the proper angle on the bezel edge in a uniform manner. The many templates aavilable will readily enable the amateur to produce a uniform shape, and a standard millimeter size. The more skilled cutter may not require the aid of a template, and produce a uniformly cut gem.

The edge of the stone should not be cut too thin as this will render it fragile, as this portion of the cabochon or facet style, will be subjected to some pressure when the metal bezel or prongs are bent over the girdle of the stone. If the girdle angle on the cabochon is too steep, this will mean added labor in difficulty in making a neat job of bending the metal bezel. If the girdle angle is not great enough the bezel will fail to clamp the gem securely and the stone is likely to work loose with future wear. In this case the cabochon will rattle in the mounting and eventually fall out—not the fault of the manufacturing jeweler.

Too much metal on the mounting bezel is likely to cover too much of the stone, and not necessarily add strength. The idea is to show as much of the gem as possible, a great exposure of metal is unsightly and not required. The tools required in setting a stone into a bezel mounting are few in number and of simple design, and readily available from most supply houses.

In setting a cabochon into a cast mounting make sure the seat upon which the stone rests is smooth and free of high spots or roughness. The ready made, stamped mountings, have smooth and even seats. The metal bezel is burnished and clamped down smooth over the stone. A small lump of beeswax is handy in handling the cabochon while fitting and adjusting the seat and bezel of the mounting.

A few remarks pertinent to the back of the cabochon may be in order. We often see a cabochon with saw marks on the back, or underside. While in the case of opaque material like turquoise and tiger eye, the back does not show when worn set in a mounting. But keep in mind that one perfectly finished stone is worth more than a dozen poorly finished examples. If the material is worth cutting and polishing at all it should be worth the added labor to sand and polish the back. Cabochons of agate and all similar transparent or translucent materials should always be polished on the back. A polished surface will admit more light to pass into the stone, and will give added "life" to the gem.

Setting cabochons in mountings is not at all difficult and with a little experience most amateurs can do a neat job. Cast mountings or ready made stamped mountings in gold and silver are readily available from supply houses. With a few tools the job can be finished, and done neatly, provided your stones are properly shaped as outlined here.

Standard Cabochons

Many home gem cutters throughout the country are building up large, magnificent, and often valuable collections of cabochons, some having as many as 5,000 or more superb and colorful stones on exhibit. In most cases these have not been cut to standard millimeter (mm.) sizes, and they are intended mainly as collectors' pieces, fancy, odd, unusual stones, and the like.

Several manufacturers make templates with a wide variety of stone shapes and sizes. This is one of four made by Bitner's. All are different.

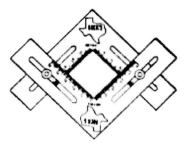

This is an adjustable template for squares and rectangles made by Rock's Lapidary Equipment. The firm also makes double templates so that matching lines can be traced on both sides of the slab.

While the hobbyist gem cutter is not primarily concerned with the commercial value or possibilities of his cabochon collection, yet the day may come when adversity, illness, and what not may overtake us. Under these conditions the home gem cutter may be obliged to resort to the sale of his gem collection. If these fine stones have not been cut to millimeter sizes, the collection will have far less value compared to the same collection in standard millimeter sizes.

The standard millimeter sizes will fit the standard ready made mountings, while odd sizes will require custom built mountings, or require the cabochons to be recut. A valuable cabochon is worthy of the added cost of a custom built mounting, but the run of the mill cabochons are not worth this added cost from a commercial standpoint. Ready made mountings cost far less than made-to-order custom built mountings.

Hence, under a forced sale, or a sale through an estate, a huge collection of cabochons would be first appraised in value in their availability from a commercial standpoint. If the gems are standard millimeter sizes the total value and sales possibilities would be greatly enhanced. On the other hand if they are not cut standard, the whole collection would have comparatively little value to the commercial buyer. It would then be a case of selling to some other gem collector, who would largely value the stones from a collector's standpoint. But in either case standard size pieces have the greatest value.

Attention should also be given to the proper slope of the edges or bezel of the cabochon. If this angle is not correct, recutting may be indicated in the event the stone is placed in a mounting. Altering the bezel angle would be a lesser problem than an entire recut to reduce to the nearest standard size.

Standard millimeter sizes (mm. abbreviation) apply to various shapes including round, oval, square, rectangular, etc. Practically all supply houses carry templates with standard sizes and shapes, intended for the convenience of the cutter. Not much more time will be consumed in using these stone size and shape guides when cutting. Not only will they give the nearest standard size, but the use of a template will also eliminate ill-shaped stones. The home gem cutter is often inclined to mark out a size and shape on a sawed slab to include the best and most desired portion for the finished cabochon.

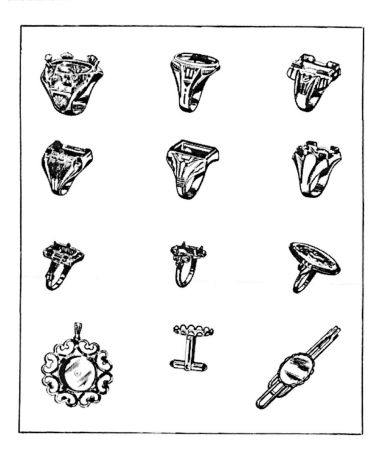

These pictures show just a few of the great variety of mountings available to the gem cutter for mounting standard sized stones. Mounting the stone is simple and easy and no soldering is required.

This can still be done in cutting to standard. Simply outline the part desired, and then fit over this the nearest standard millimeter size, and cut accordingly. After the blank is cut out on the diamond saw, the template is used from time to time during the grinding operation. In the grinding operation where the stone is cut to shape, about 0.5 millimeter (25 millimeters equal one inch) excess should be left on outline, as the sanding and polishing operations will remove approximately 0.5 millimeter of material. All this may sound complicated and involved, but it is not so. A little experience will soon develop skill of size and shape judgment. The jump from one standard millimeter mounting size to the next larger or smaller is not great. Standard mm. sizes are used in both cabochon and facet cut stones.

Polishing Jade

Numerous methods have been given for the sanding and polishing of jade, for example. Some claim they can only attain a satisfactory polish on jade by the use of a leather polishing wheel. As a mater of fact it is possible to obtain the fine polish on jade by the use of the standard "rock hard" felt buff.

The writer watched the late Oscar Smith of Portland many times quickly and seemingly easily put a fine polish on jade using standard dry sanding methods, and polishing on the rock hard felt buff with common tripoli(fine grade) as the polishing agent. Smith had been a commercial gem cutter for some 30 years. He was enabled to polish practically every cabochon stone on the felt buff with tripoli. The only time tin oxide or similar costly polishing powders were used was when dealing with a valuable opal, when tin oxide and a medium hard felt buff was used to follow the tripoli to attain a final "gloss" finish.

We have seen many gem cutters both amateur and professional use a lapidary technic that was not generally regarded as being standard or orthodox, yet they would attain satisfactory final results. The whole point to remember in lapidary work is that the technic employed may not be as important as the skill and experience one may have in a given method. In short, an artisan may use an unorthodox technic for a given task, yet through long experience with same he is enabled to get equally good final results compared to the worker using what may generally be regarded as the proper and "correct" methods.

Coming back to the polishing of jade on the felt buff with tripoli, the whole "secret" seemed to be using heavy pressure, and allowing the stone to run dry and hot in the final polishing stages. In fact Smith let the stone heat up to a point just under the melting point of the dopping wax. This seemed to produce that much sought for molecular surface "flow" which yields that fine glossy finish. It is next to impossible to describe this method in print; it is wholly a matter of the operator gaining the skill through experience. Smith was inclined to look askance at the leather buff, but doubtless the less skilled gem cutter will find the leather buff best for items like jade.

Regardless of what technic you may be using in various types of lapidary work, including facet cutting, it is quite probable that with enough experience you will gain enough skill in your method to sooner or later be enabled to do equally good work as compared to any other method. A good example may be seen with the apprentice mechanic in any field; he is likely to appear on the job with a trunk full of tools of various kinds. The experienced worker will come along and perhaps do better work with only one-tenth of the tools—wholly a matter of experience and skill.

Owing to the fibrous structure of jade the amateur lacking experience with this material may sometimes find it difficult to obtain a high glossy polish. This is especially true when working with large flat surfaces. The curved surface of the relatively much smaller cabochon is much easier to finish compared to a large flat surface. This same theory is applicable to any gem material.

In working the Wyoming nephrite variety of jade, Allan Branham suggests the following technic:

(1) Rough grind with 100-grit silicon carbide wheel or the loose grit when working on the horizontal lap. (2) Finish grinding with 180-grit. (3) Sand on new dry sander of 220-

grit. (4) Finish sanding on well-worn sander of same original grit. (5) Polish on leather lap with tin oxide. Branham does not advise the use of a felt buff in the final polish.

Various technics have been suggested for polishing of tough gem materials like jade. Some lapidarists report good results by the use of the "rock hard" felt buff, tripoli, and heavy pressure during final polishing. The main point to remember is to remove all scratches from the surface at the sanding operation before passing the final finishing.

The tendency is often to hurry the work past the sanders. This only means failure in obtaining a satisfactory high polish, or much time must be spent in the final polishing operation. Remember that the sanders will remove deep scratches much faster than the leather or felt or any other buff which may be used in the final polishing operation.

The Fibrous Gems

In finishing cabochons and flat surfaces of gem materials having a distinctly fibrous structure, the sanding and final polishing should always be done at right angles to the direction in which the fibres rest. This is especially true with a gem material like tiger eye, it being next to impossible to attain a proper final polish if this technic is not followed.

There are other gems where some difficulty may be encountered with the fibrous structure, but perhaps none as prominent as tiger eye. Other fibrous species include chrysoberyl (cat's eye), jade, tourmaline, agate, malachite, and a few others. In most of these species the fibrous structure may not be noted by the unaided eye, but may be clearly revealed under magnification. If difficulty is encountered in polishing any of these gem materials, try polishing the cabochon or flat surface from one direction only. By experimenting from different directions the correct angle direction will be found.

Difficulty may also be encountered in polishing a large flat surface like the table on certain facet cut gems. This difficulty, however, is usually not due to any fibrous structure, but a case of sloughage in dealing with a material having a marked cleavage in one direction. This difficulty may often be overcome by redopping the gem and polishing from the opposite direction. In the case of topaz, with a very marked basal cleavage, a change of direction will not aid in eliminating sloughing during the final polishing. Topaz should be so oriented that the angle will be about 5 to 8 degrees off the basal cleavage, and not exactly parallel with same.

If difficulty is encountered in polishing a large table on a facet gem there is one final and usually quite effective manner of dealing with it. Simply go to the felt buff and in a matter of minutes an excellent polish will be attained. This, however, is not regarded as skilled facet cutting, for it will be noted by the trained eye that where the crown table facets meet with the table the edges are not sharp as they would or should be normally, but will appear slightly rounded.

CUTTING ASTERIATED QUARTZ CABOCHONS

MERWYN BLY

The production of a truly symmetrical and pleasing finished gem with a lively, well-centered star, requires only a minimum of skill and optical equipment if certain technics are followed. Each cutter has his own method. The writer is describing his in some detail, and with some background, in the hope that it may be of value to those who have established no technics of their own.

A basic understanding of the cause for asterism in quartz is considered helpful. As is known quartz is uniaxial, hexagonal. The "c" or optic axis is parallel to the length of the normal crystal. The characteristic six rayed star is exhibited by material so cut that the optic axis passes vertically through the center of the top and bottom of the cabochon. Off-center cutting will produce off-center stars. Serious errors may produce random length rays or less than six rays.

According to Albert J. Walcott, asterism in quartz is an edge effect from included striations formed by an oscillatory growth of two or more forms on one crystal. The striae form three great circle chatoyant bands intersecting at angles of 60 degrees at the pole of the "c" axis. Three intersecting bands thus cause a six-rayed star. This is really a diffraction effect similar to the "star" seen by looking at a distant street light against a dark background through a fine mesh wire screen.

So much for a hasty outline of the cause of asterism. Asterated quartz is probably more common than we think. Those rose varieties, most suitable for the purpose, come from Brazil, Maine, and South Dakota. It is almost invariably badly fractured which represents the only real difficulty in cutting it.

After choosing an area as free from flaws and fractures as possible, saw out a rough cube somewhat larger than twice the size of the finished cabochons desired. Grind off the corners of the cube and work it down on the wheel into a rough sphere. It is not necessary that a true sphere be produced, only that all sharp projections or flats be removed to the extent that from any aspect a reasonably rounded sphere be presented. Now sand the rough sphere and give it a superficial polish; polish enough so that when wet with oil or water no opaque areas remain. Waste no time attempting to get a polish necessary to pick up the star by reflected light. Merely wet your semi-polished approximate sphere and look through it at a small electric light. A "star" of some sort should be apparent no matter what the orientation.

Four rays with acute and obtuse angles indicates an orientation approximately 90 degrees from the desired "c" axis. Keep the stone wet and continue to turn it about until a six-rayed star is seen; now turn it very slightly this way and that, until the arms and ray angles appear equal and the apex (center of the star) is as small and as well defined as possible. Definition will not be too good at this stage but more than sufficient to mark the spot where the apex appears on the stone. This point represents one pole of the optic axis. Turn the stone through 180 degrees and view the star through the end opposite that already marked and mark this apex. The two marks now represent the opposite poles of the optic "c" axis. Join them by a circle passing clear around the stone; draw a circle at the "equator" at right angles and you have a mark for sawing in half.

When sawed, the two halves represent two roughed-out cabochons. Using the now flat bottoms as your guide planes, true them up, sand and polish, and you should come out with a pair of star quartz gems with reasonably well-centered stars.

Assuming the gem has been centered to the satisfaction of the cutter and brought to the highest possible polish, both top and bottom, the question of backing arises. Very dark rose quartz requires no backing and the bottom may be left unpolished, like star sapphire, it is said. Unfortunately, the writer has never been able to secure such material. Ordinary rose quartz in direct sunlight exhibits a brilliant and attractive star, as does clear asterated quartz to a lesser degree. However, for use in jewelry or in collections, without special point sources of light, it is awkward and unsatisfactory to have to explain the star rather than exhibit it, so some sort of backing is in order.

The time-honored method of backing is to grind and glue on pieces of colored mirror, blue or red, to simulate asterated corundum. Speaking only for himself, the writer doesn't care for the system. Star quartz can stand on its own merits without attempting to imitate something else. Star quartz, the writer feels, should be backed as matter of factly as fire opal, and as openly.

One very effective method is to have metal, preferably gold, "sputtered" on the back of the gem to bring out the star. If the stone is set "deep" and a protective lacquer brushed over the gold surface seems to resist all but actual abrasion very well. The effect is undeniably pleasing. However, it requires access to a laboratory or one of the commercial concerns who ordinarily "sputter" mirrors for amateur astronomers.

The "Star of Anakie," a huge star sapphire cut and owned by Kazanjian Brothers, Los Angeles.

The second method is to have a bezel fitted with a flat bottom plate of gold and merely set the stone so it rests on the plate, no glue or balsam being used. The writer has found this surprisingly satisfactory. A variation is to use a bit of chromium plate, polished brass, aluminum, in fact any bright metal free from tarnish, or treated to be so. By this means all the disadvantages of Canada balsam stuck backing is avoided, and as far as the writer can see the star shows up just as well as though the backing were glued.

Cutting Star Sapphire

Asterated sapphire must be cut cabochon style and since this material has a high hardness some special problems in sanding and polishing are presented. The rough pebble or crystal must first be oriented in relation to the crystal axis. The star will appear best only when the rounded cabochon surface is in the plant of the "a" axis, and at right angles to the principle "c" axis.

A sapphire crystal is often elongated and prismatic in outline, with the "c" axis running lengthwise through the crystal. The rough material can be shaped to form and size on a silicon carbide grinding wheel. Generally the back of the cabochon is left slightly convex and is left smooth but usually not polished.

All opaque or slightly opaque sapphires are not necessarily asterated, and there appears to be no external indications of the possible presence of a star. Actual cutting seems to be the most effective means of revealing possible asterism.

To remove scratches left by the grinding wheels, vertical running sanding wheels are used. Cloth charged with 220 grit silicon carbide are satisfactory for the sanding operation. Polishing of the gem can be accomplished on a leather buff, which will prove quite slow. The lead lap will polish star sapphire much faster and is the method used in many commercial lapidary establishments. Various standard polishing agents may be used including tripoli, alumina, and buffing powders. Very fine grit *Norbide* is also of value in obtaining a final polish. Regardless of the polishing agent used the final polishing of star sapphire will prove relatively slow.

Cutting Labradorite

Some gem cutters, especially those not having had previous experience, often have difficulty in the lapidary treatment of labradorite. This colorful gem material is best suited for cabinet specimens, cut as a flat surface and polished. Often the play of beautiful peacock-like colors in this material is remarkable.

It is also possible to cut fine cabochons from this material, but there is likely to be more waste compared to flat surfaces. Since labradorite (H-5 to 6) is comparatively soft the gem would not be suitable for ring wear.

A play of colors is characteristic with certain types of labradorite, especially those from Canada. Blue and green are the predominant colors, but yellow, fire-red and pearl-gray may also be noted. This effect has been shown to be largely due to the interference of light, caused by reflection from thin lamellar inclusions of various minerals, according to Dana. These inclusions are arranged in parallel manner and in certain planes in the material.

Frequently this plane, showing the best play of colors, may be noted on a natural cleavage surface. In any case, in sawing the material for a polished surface specimen, the specimen at hand should be studied in order to locate these planes of best color. Often the sawed surface may not reveal the best color, in which case it will be necessary to reduce the surface by grinding on a horizontal lap wheel, or on the side of a grinding wheel, examining the surface at frequent intervals to note when a good exposure of color is made.

This orientation of the material is not so important in a curved cabochon surface, for here a number of planes of color will be exposed, but the general planes of color must be parallel with the flat base of the cabochon. The proper orientation of labradorite is a matter of close observation and experience.

To simply take a rough mass of the material and proceed to section it is only likely to give good results through sheer good luck. A good rule to follow is to saw or grind parallel to the largest area of natural cleavage, and this is usually seen on most specimens.

In cutting labradorite, the material is a good deal like the familiar white fire opal of Australia. Every cutter who has dealt with Australian opal soon learns that the play of colors runs in layers or seams, and these are generally parallel to one another and often close together. Hence an experienced cuter of opal will study the rough piece and from the sides note where the best color rests, and then either saw acordingly, or carefully grind until the best seam is exposed. If this type of opal is carelessly cut at right angles to these "fire" layers or seams, the finished stone is likely to be practically worthless. This same applies also to lauraaorne.

Like in opal, the best colored layers are not very thick, hence due care must be used in grinding away just enough, not too much and not too little. Labradorite, being one of the feldspar group of minerals, is quite soft, it saws and grinds very easily, being easy on the saw as well as on the grinding wheels.

By far the best gem quality of labradorite comes from eastern Canada, including Labrador, Greenland; some is found in Ontario and Quebec. The best grade has been found in eastern Labrador, In the past, Labradorite has never been plentiful in the United States.

Small lots would be imported from time to time by some supply firms, and when the supply was sold, no more might be available for a long time. A good many of these importations came by boat, either from Greenland or from Labrador, and usually only a ton or so at one shipment, the price was never low.

Buying the Rough

Certain types of rough gem cutting materials are often offered at so much per pound or piece. Obviously, under these circumstances, with many gem materials a certain element of speculation will enter into the transaction. Some items sold under designation of "mine run" mean just as it is gathered in the field, from stream gravels, quarry, or from a hole dug in the ground. Then there are various other specific designations including choice, selected, well crystallized, crystals, and many others.

All of these terms mean something, but no matter how honestly the seller may "select," it should be kept in mind that he generally cannot see inside of the rough mass. Certain types of material from specific and well known localities will average to certain grades and this is fully understood by those who are experienced in handling this material. The more common and cheaper grades of rough gem materials, sold by the pound, will in the long run average out profitably when cut and polished. However, it is possible that in a single pound or piece, selected as mine run or at random, may prove to be poor quality or unsuitable for cutting. This could hardly happen in a larger similarly selected lot.

For many years it was customary for the Australian opal miner to sell his production of rougn opal to the buyers visiting the diggings. Plenty of speculation entered into these transactions. At times the buyer charged up a loss, or the seller obtained either more or less than the true value of the material. But neither the buyer or the seller was looked upon as a shady character. The only really accurate means of estimating the value of a lot of rough opal would be to reduce all of it to finished cut gems—and this was not practical, or was simply not being done. Both parties took the speculative chances, but at the same time tried to rely upon their good judgment and past experience.

Obviously in a transaction involving rough gem cutting materials of some types if the buyer is on the short end it is only human that he may be inclined to regard the seller with more or less suspicion. Or if the seller learns that he has sold something for only a small part of its "internal" value he is not inclined to feel very happy over the deal.

While we get very few complaints regarding the sale of rough gem materials, the few that have been received are generally not justified upon investigation. In many instances we learn that the complainant is a novice, and not familiar with values. Or the novice may lack experience and judgment on the proper manner in which gem materials should be cut to the best advantage. Very rarely do we learn of deliberate fraud on the part of the seller.

Sawing the rough material into sections suitable for cabochon cutting or polished cabinet specimens is generally a reliable means of proving the quality of the material at hand. With a pencil outline the number and size of the stones which can be cut may be readily determined, and this in turn is an excellent manner in which a fairly accurate value may be arrived at.

The finer and more valuable types of cabochon cutting

materials are generally sold in the form of sawed slabs or sections, a method which is fair to both the seller and the buyer. The value of the slab may be arrived at by the number, size, and quality of gems which the section can produce.

Agate-filled nodules, thunder-eggs, geodes, concretions, and similar items are sometimes sold sight unseen. It is usually next to impossible to determine just what the interior may yield. Specimens of this kind from most localities will usually average along certain percentages between worthless, poor, medium, and high grade. This should be understood by the buyer. In some cases it may be a good average to find only ten per cent in the high grade class and possibly only one per cent in the superfine group.

In the old days moss agate was gathered up along the Yellowstone River in Montana by the ton. This was generally carried on as a small local industry, and the "mine run" brought about ten cents a pound. The buyer would perhaps learn that fifty per cent or more would prove next to worthless for gem cutting, and only a small percentage really high grade scenic agate. But it did not require many of these high grade specimens in a ton to compensate the buyer for the duds. Mine run agate of this kind now brings about ten times as much as it did at one time, and with the growing scarcity of this material the top price is not yet in sight.

In selling mine run the seller could sort over the material and pass along only duds, but this fraud would be open to plenty of suspicion if only duds were found in a reasonable size lot.

In buying a few pounds of rough gem cutting material, costing perhaps a few dollars, you may cut gems from this material worth many times its original value. On the other hand the next or another lot may not be worth the cost of cutting. In the long run it has been our experience that buying in the rough is not at all fraught with any great or continued loss—averages will prove profitable even when taken as mine run.

The Coloring of Agates German Method

The German agate cutting industry has long practiced artificial coloring of agates. As a rule this "beauty bafh" treatment of agate can be instantly detected by the expert familiar with agate. The below data on agate treatment is from a paper published originally in Germany in 1913 by Dr. O. Dreher of the Idar agate cutting center. The information as given in the original paper by Dr. Dreher was more or less in the nature of a trade secret. Practically all of the English papers published on this subject have been translated from the original work of Dr. Dreher, as given below.

The coloring of agates depends on the introduction of a coloring matter into their pores. Some layers of agate are less porous and therefore these will not absorb pigments, but remain wholly uncolored or only partially colored. The cutter calls the less porous layers in the agate *hard*. The layers or bands readily colored are termed *soft*. The skilled artisan can often judge the ability of an agate to absorb pigments, prior to the treatment.

The art of coloring agates and similar stones has been known to us for only a relatively short time. Long ago the Romans had learned the secret of the black colors, but they kept this secret for centuries. Finally in 1819 this old Roman technic was discovered by accident.

Lessons from Nature

Along about 1813 some German cutters observed agates in the field, presumably colored by the action of sunlight. Agates which projected from the earth were often colored a carnelian or sard (reddish), while the remainder of the stone beneath would be entirely colorless. This led to the practice of "burning" colorless agates to produce the reddish colors.

Not all colorless agates will become reddish when given the heat treatment of "burning." It is thought that the agates which fail to respond are those lacking in iron compounds, present as an impurity. This was finally solved by soaking the agate in a soluble iron salt and then "burning" by oven treatment. In 1845 the method of blue coloring was discovered and in 1853 green colorings were introduced, all the result of experiment by the lapidarist.

Different Methods

The manner in which the coloring pigment is introduced into the agate varies according to the color desired. In all cases where a permanent color is attained, the coloring matter is not introduced in a dissolved form directly, but by the use of various chemical reactions; these take place within the agate.

In general there are two methods of coloring an agate. In one case the soluble metallic salt is permitted to soak into the pores of the agate. This soluble salt in turn is changed to a colored insoluble oxide by heating. In the other method two solutions or "baths" are used in succession, the second bath causing a colored precipitate of an insoluble metallic salt to be deposited within the agate.

The following will serve to illustrate how some of the colors can be obtained in an agate:

Red—Soaking stone in iron nitrate solution and then by "burning" an iron oxide is produced.

Bluish Green—Soaking in solution of chromic acid or ammonium bichromate, and heating to produce a chromic oxide.

Apple Green—Soak in nickel nitrate and "burn" to produce a nickel oxide.

Brown—Soak in a sugar solution and heat strongly to

carbonize to caramel.

Blue—Soak in bath of yellow prussiate of potassium and then in a solution of iron sulphate to precipitate "Berlin blue."

Blue—Soak in solution of red prussiate of potassium and then in solution of iron sulphate to precipitate. "Turn-bull blue" in agate.

Black—Soak in sugar solution and then in sulphuric acid to change sugar to carbon.

For completeness it may be mentioned that aniline dyes have been used to some extent in the artificial coloring of agates. The aniline colors, however, are not as permanent as the metallic oxides and precipitates described above. Aniline tends to fade when exposed to strong light.

Extraction

Before the agate is colored it must be cleaned of all oil and impurities which may be adhering to or soaked into the stone. In the cutting of agates, oil or kerosene is used to lubricate the saws, and this must be first "extracted." The petroleum substances can be removed by boiling in a strong solution of sodium bicarbonate, or solvents like gasoline or some non-inflammable commercial cleaning fluid can be used "cold."

The agate may carry a small amount of iron and it is desired to remove this prior to "burning" for green colors, otherwise a dull or muddy green may be obtained. To remove iron compounds the stone is placed in warm nitric acid for two or three days and then placed in warm water for several days. The purpose of the nitric acid is to render any iron present soluble, so the water soaking may remove same. The warm water bath should be changed a number of times.

Important Colors—Red

The knowledge of obtaining carnelian and sard (reddish) colors in agate by "dry burning" was first discovered in 1813, but the "bath" method of obtaining the red shades came later, and at uncertain date.

Red—To produce *red* colors the agate is soaked in a strong solution of iron nitrate. According to the directions of the old German agate cutters, this solution should be as thin as Munich beer. The aquaeous solution of iron nitrate should be kept warm and the agate submerged for from one to four weeks according to the thickness of the stone.

Stones three mm. thick for about a week, six mm. about three weeks, and ten mm. stones about four weeks. Stones thicker than ten mm. will seldom color throughout. (A mm. is 1/25 of an inch.) This means that seldom will the color penetrate into an agate deeper than about five mm., or about 1/5th of an inch. Let it be understood at this point that all coloring is done *after* the stone or slab is completely cut and polished, otherwise grinding would expose the uncolored material below.

A section of fine Australian agate.

Drying Agate

After the agate has been soaked in the above solution for the desired time it should first be carefully dried in a warm oven for from two to ten days. This is to remove as much free moisture as possible prior to the "burning" to avoid possible fracturing.

Burning

The agate is removed from the oven and while still warm is placed in a crucible. The agates can be best packed in some substance like fibrous asbestos or powdered magnesium oxide, and the crucible cover (an iron crucible will answer).

The heat in the oven is raised very slowly, until the crucible has reached a red heat. It is then allowed to cool very slowly. This is best carried out by reducing the flame or heat gradually. The agate must not be removed from the crucible until the contents are completely cooled.

It is possible that some stones may not have the desired color. In this case the soaking in the iron nitrate solution and the oven "burning" can be repeated one or more times as desired.

Green

Green colors can be produced in a number of ways. Two "baths" are in common use—saturated or strong solutions of chromic acid or potassium bichromate. The solution of chromic acid seems to be preferred, although the bichromate salt is cheaper.

The stone is placed in the chromic acid solution for from eight to fourteen days, according to the thickness and the "hardness" of the agate. Stones or slabs over ten mm. in thickness should remain in the bath for a longer time, up to eight weeks.

The stones are then removed from the bath and placed in a warm closed container with lumps of ammonium carbonate, for at least two weeks. The purpose here is to have the ammonia gas penetrate the agate and cause a bright green precipitate of a chromate salt. (Liquid ammonia solution would possibly bleach out some of the soluble chromic acid or bichromate.) After the agate is removed from the ammonia gas chamber it is dried and then gradually strongly heated in a crucible and oven as described under red coloring.

Green colors often do not come up to expectations. A muddy green or bluish-green may be noted. Experiments will often solve the problem in various kinds of agate.

Black Carbon

Black coloring was first known to the Idar cutters in 1819, and was discovered in an accidental manner.

The agate is first soaked in a solution of ordinary sugar, 375 grams to one liter of water, or about as thick as flowing honey. The earlier cutters employed diluted honey, hence this solution is often called the *honey bath*. While the agate is in the sugar solution the vessel should be kept warm, as this seems to promote penetration. The stone is kept submerged for from one week to three weeks, according to thicknss, "hardness," and depth of color desired. As water evaporates from the warm solution additional water can be added.

The agate is removed from the sugar solution and without washing is placed in sulphuric acid. The acid is slowly warmed on a hot plate and then brought to a boiling or near the boiling point for about fifteen minutes. The vessel should be covered and care should be exercised to avoid the hot acid from spattering in the eyes, skin or clothing. A large vessel is best and a hot plate where the heat can be controlled is excellent. The agate is permitted to cool with the acid for a few hours.

Blue

Blue coloring was first used at Idar in 1845. Two shades of blue can be had by the use of yellow prussiate of potassium or by the use of the red prussiate of potassium (ferrocyanides of potassium).

Dissolve 250 grams of one of the above salts (poisonous) in one liter of water. The agate is soaked in this solution for from one week to two weeks. This bath should be kept warm, but not too hot and should not be boiled.

The agate is then soaked in a solution of iron vitriol (iron sulphate) for from four to eight days according to the depth of color desired. No "burning" is needed in this method.

A darker blue color will be had if the iron sulphate solution is acidified with a few drops of sulphuric and nitric acid. While the agate is in the iron sulphate solution it can be examined from time to time, and removed when the desired color is noted. The solutions used in agate coloring can be used repeatedly by adding water to replace evaporation and small amounts of the salt as the liquid becomes weakened.

Types of Equipment

"What type of lapidary equipment do you advise?" This is a frequent question we have been asked by those who contemplate taking up lapidary work as a hobby. As a matter of fact this exact question has been put to us hundreds and hundreds of times during the past 25 years.

The obvious answer to this question would be "What are your requirements?" Attention is called to the fact that it is possible to do equally good work with a small and modestly priced unit, as it is with units in the much higher priced field. We generally refer to the higher priced units as being in the commercial type class, and suitable for mass production.

In short, the home gem cutter with a quite modestly priced lapidary unit will work say two or three hours, and perhaps finish only two or three cabochons or polished pieces. Putting in the same time on the higher priced "commercial" type units, the same operator will finish from 10 to 20 times as many cabochons or polished pieces. But it is to be remembered that in both instances the work turned out can be equally good in quality.

The difference in production rests in the fact that in the commercial type units, two operations may be carried out at the same time. While a specimen is in the saw, the operator can be working at sanding, grinding or polishing. In the lower priced type of equipment, it is generally necessary to change tools for each operation, and this takes added time.

So in deciding what type of equipment to purchase at the outset, it may be well to keep these facts in mind. Or better yet, visit a number of home lapidary shops, talk to those who have had experience, note the equipment used, note the many type of excellent factory built equipment available, and then decide what would best suit your needs, and how much money you care to invest in your hobby at the outset.

Many do start out with low priced units, and continue with same for years, while others decide to "graduate" into mass production equipment. Attention is called to the fact that it is generally possible to readily dispose of any standard, well known piece of used lapidary equipment. Standard lapidary equipment always has had, and still has, a good resale value. Note in the advertisements in the magazines in this field how seldom used lapidary equipment is offered. It is usually possible to dispose of it privately without being obliged to advertise. Hence no matter what type of standard, factory built equipment you may purchase, it has a definitely good resale value, provided same is maintained in good condition—just like your Ford or Chev.

It may be noted here that while the hobbyist is not primarily interested in mass production or commercial work, the fact is that many of the old timers who engage in gem cutting, easily make the hobby pay its own way, and at the same time build up valuable collections—all in leisure time. Finished gems, and polished specimens always have a ready market.

The home gem cutter is bound to finish many duplicates which are not needed for his private collection. For example, when a thunder egg is sawed there will be two halves, more or less identical or quite similar. The cutter will generally place the best half in his collection, and the other half becomes a duplicate not needed. The same is true in finishing many other specimens, including cabochons. Hence the home gem cutter is bound to accumulate duplicates. These may be disposed of in quantity lots to supply houses, or even to beginners in the field. When sold to supply houses, retail prices are not to be expected, for after all the dealer must realize costs of operation and profit.

The income from duplicates and surplus production enables many hobbyists to make the hobby pay its way, and at the same time, over a period of years, the home cutter may build up magnificent collections valued in the thousands of dollars. We have seen this done many times during the past 25 years. There are few hobbies that can be made to pay their own way.

Adapting Sanding Cloth

Wrinkles may appear when adapting the sanding cloth over the working face of the vertical running type of sander. In order to get a smooth adaptation of a new sanding cloth, the back side (not abrasive side) should be made slightly moist with water. This will soften the cloth and permit easier adaptation to the face of the wooden disk.

On this type of sander the cloth is generally held firmly in place by the use of a metal hoop, fitted tightly over the slightly tapered periphery of the wheel. Some workers cut a groove on the periphery of the disk, and hold the sanding cloth in position with a heavy rubber band cut from an old inner tube. In both cases the dampened cloth can be more easily adapted.

USEFUL LAPIDARY NOTES

How to Cut Star Quartz

Certain types of quartz show asterism on the top of the curved cabochon when the material is properly oriented in relation to the optic axis of the material. Quartz crystallizes in the hexagonal system, hence it has but one isotropic direction. This direction is parallel to the vertical crystal (c axis) axis. To show a star the top of the cabochon must be cut at right angles to the c axis. Some types of quartz will yield a bold six-rayed star when properly cut. Other material when oriented at right angles to the axis or when improperly oriented may show only a four-rayed star.

Material Used

There are a number of localities which produce rose quartz suitable for cutting asterated cabochons. Much of the rose quartz from South American is almost colorless—having only a faint trace of pink. Most of the rose quartz from American localities is less translucent and more pink in color. The pink American rose quartz may show a conspicuous star. The cabochon should not be polished on the back and the sides of the cabochon should be cut rather steep. Rose quartz which presents a milky appearance will often show the best star. The nearly colorless material from South America requires backing with a thin colored mirror in order to reflect the star through the top of the cabochon.

Orientation

It is necessary to locate the optic axes on the material before cutting. The most simple manner in which this may be accomplished without optical aids is to first shape the material into a sphere and polish the surface. By a careful examination of the surface of the sphere it will be possible to note where the six-rayed star appears to the best advantage. The sphere is then sawed into two parts, and each half shaped into a cabochon stone.

A dichroscope or two sheets of *Polaroid* may also be used to orient the quartz optically. When rose quartz is viewed through a polariscope and rotated, looking through the a axes, there will be four alternate positions of light and dark. The quartz may also be held between two sheets of *Polaroid*, and the speciment rotated while held toward a strong light, when the positions of light and dark will appear. The sheets of *Polaroid*, while being used to view the specimen, should be in a position where the least amount of light will be passed. This corresponds to the "crossed nicols" on a petrographic microscope. In viewing rose quartz between sheets of *Polaroid*, the positions in which the specimen appears at its greatest darkness are the "positions of extinction," of which there will be four when the specimen is rotated through a complete circle.

Other optical instruments may be used to orient the quartz. A thin section may be sawed from the specimen and examined under the petrographic microscope, or an ordinary low power microscope fitted with *Polaroid* attachments. The hand polariscope will also serve to orient the material.

Cutting Gem

After the material has been properly oriented, cut the gem in the usual cabochon style. Best results will be obtained from a rather high oval rather than a flat stone. Where the material is fairly dark no backing will be required, in which case shape the bottom of the cabochon slightly oval, and do not polish the bottom. The cabochon sides should be well polished.

"Star Sapphire"

The nearly colorless asterated rose quartz from South America must be backed with a reflecting surface of some kind to bring out the star. Material of this kind is usually backed with a thin blue colored section of a mirror, and a "doublet" of this kind may exhibit a splendid blue color and a conspicuous six-rayed star. Stones of this kind are sold as imitation star sapphires.

In backing with a mirror, the bottom of the cabochon is left flat, while the sides are cut steep in the usual manner. A little experience will indicate the steepness of the sides to bring out the star to the best advantage. Inexpensive and thin blue colored mirror may be used as a backing material to bring out the star. Commercial blue colored mirror may be too thick, but this can be easily remedied by first sawing the mirror into small sections. The mirror is then ground down on the side of a fine grit grinding wheel or lapped down on a horizontal running lap wheel. The mirror section used for backing should be ground down to about 1/32 inch in thickness. In grinding the mirror care should be taken to avoid disturbing the backing of the mirror. After grinding the mirror, the surface need not be repolished, but it should be sanded or lapped smooth and flat. The cementing material will eliminate the necessity of repolishing the mirror.

The mirror backing when cemented into position may be left oversize, the projecting edges can be ground off flush and polished after the cement has set. The mirror backing may be cemented to the bottom of the cabochon with a number of cements which are wholly transparent. Canada balsam has been found very satisfactory as a cement. Apply a large drop of the Canada balsam to the mirror surface and press down the bottom of the cabochon. To hasten the drying of the Canada balsam the mounted "doublet" may be placed on a warming plate or in an oven and heated for about ten minutes to about 180 degrees. Overheating may cause the balsam to boil and disturb the position of the two parts. After heating allow to cool for several hours. The balsam cement may be loosened at any time by heating the stone to about the boiling point of water.

Some of the skilled commercial gem cutters and those who have had considerable experience in handling asterated quartz can generally orient rose quartz satisfactorily by careful examination with the unaided eye. The novice, however, will need some optical aids, or cut spheres, or simply use "guess" work.

In orienting garnet crystals for possible asterism, it is generally customary to round off and partly polish several areas and examine for a star. If the partly polished surface is moistened with olive oil the star, if present, will be better revealed. A single small flashlight is best to examine the surface for the presence of a possible star.

In examining a sphere of nearly colorless quartz for the possible presence of a star, the polished sphere should be held toward a strong single light, and the sphere rotated. The more highly colored and more opaque rose quartz will of course reveal the star by a surface examination.

As a final finish to the mirror-backed cabochon, the mirror

may be coated with a black or some dark colored quick drying lacquer for protection from scratching.

CUTTING STAR SAPPHIRE

Asterated sapphire must be cut cabochon style, and since this material has a high hardness, some special problems in sanding and polishing are presented. The rough pebble or crystal must first be oriented in relation to the crystal axis. The star will appear best only when the rounded cabochon surface is in the plane of the *"a"* axis, and at right angles to the principle *"c"* axis.

A sapphire crystal is often elongated and prismatic in outline, with the *"c"* axis running lengthwise through the crystal. The rough material can be shaped to form and size on a silicon carbide grinding wheel. Generally the back of the cabochon is left slightly convex, and is left smooth, but usually not polished.

All opaque or slightly opaque sapphires are not necessarily asterated, and there appears to be no external indications of the possible presence of a star. Actual cutting seems to be the most effective means of revealing possible asterism.

To remove scratches left by the grinding wheels, vertical running sanding wheels are used. Cloths charged with

220 grit silicon carbide are satisfactory for the sanding operation. Polishing of the gem can be accomplished on a leather buff, which will prove quite slow. The lead lap will polish star sapphire much faster and is the method used in many commercial lapidary establishments. Various standard polishing agents may be used, including tripoli, alumina, and buffing powders. Very fine grit *Norbide* is also of value in obtaining a final polish. Regardless of the polishing agent used, the final polishing of star sapphire will prove relatively slow.

RUBBER BONDED WHEELS

The rubber bonded, silicon carbide wheels have been tested in the Gem Cutting Laboratories of THE MINERALOGIST. Rubber bonded silicon carbide wheels will find wide utility in gem cutting, jewelry making, and metal work. Below are a number of suggestions.

Rubber bonded wheels are operated dry and at approximately the same speed as the regular silicon carbide wheel. The 6×1 inch, 100 grit wheel is well suited for general lapidary purposes. With proper care a 6-inch wheel will give very long service, since they are not used for heavy grinding, and they do not slough off rapidly.

Sanding Sapphire

Sanding the wheel marks from cabochon cut sapphire is a slow operation when carried out on the regular sanding cloths. The 100 grit, rubber bonded silicon carbide wheel will "sand" star sapphire much faster than regular sanding cloth. The rubber bonded wheel is especially efficient for this purpose.

Corner Sanding

In cutting unusual shapes like crosses, hearts, and crescents from hard gem material like agate, difficulty may be experienced in sanding the angles and corners prior to polishing. The rubber bonded wheel will be found ideal for this purpose. Marks left by the regular silicon carbide grinding wheels can be removed with ease and speed.

Cabochon Sanding

Cabochons cut from hard material can be readily sanded with the rubber bonded wheel. The rubber bonded wheel is not indicated for sanding very soft gem materials. For the sanding of cabochons the rubber bonded wheel has the added advantage in that no noticeable dust or grit is thrown from the wheel.

Metal Grinding

For finishing jewelry work after casting or soldering, prior to polishing, rubber bonded wheels find wide utility. This type of wheel is especially efficient in metal work, quickly removing scratches and leaving a smooth surface ready for the final polish. They also are clean to operate, and are available in a wide range of grits and sizes.

The side of the rubber bonded wheel can also be used for sanding flat surfaces, like the backs of cabochons. The most suitable rubber bonded silicon carbide wheels, for use in the gem cutting industry are the 60 and 100 grits.

LAPIDARY NOTES

FOR MUD SAWING—The special "B" grade of *Crystolon* is made purposely for this use. They are special grits which will adhere to the sawing blade better than ordinary silicon carbide.

TESTING AMBER—Among the most common cheap imitations of amber is Bakelite. A simple test will serve to distinguish between them. Amber is very light and will float readily in strong salt water, while Bakelite and similar substitutes will promptly "hit bottom." Hence, if the lady's strand of beads sinks in a solution of salt water, you can confidentially inform her, if you dare, that there is something wrong. The cheaper types of amber—the "pressed" and "moulded" varieties—will of course float, since they are actually the real article.

TURQUOISE—There are numerous substitutes for

turquoise and care must be exercised in passing opinion on stones of this kind. The ruby was originally made by fusing together very small fragments of genuine material to obtain a larger and more valuable gem. So are sometimes large masses of turquoise obtained, except that heat is not used, but small fragments are pressed together; and probably some cementing substance is added. Turquoise is rarely found in rough masses large enough to cut a large stone. Hence, the large, inexpensive (or expensive) gems sometimes sold may be a worked-over turquoise or an out and out imitation. Care should be exercised in permitting a fine turquoise to come in contact with oil or soapy water. The gem should occasionally be cleaned with dilute ammonia water, which also aids in improving the color.

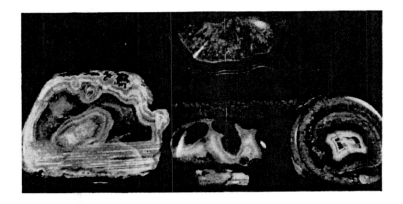

Fig. 33-B. Large polished agate specimens. (Top) Brazil, (lower) Wyoming and Black Hills of South Dakota. (Photo Rushmore Museum, Rapid City, South Dakota.)

IMITATIONS—The ingenuity of man has reached a point where a great many gem stones can be manufactured in excellent imitations, which may even pass the casual observation of an expert. Probably as many European manufactured stones are sold in the Oriental gem producing countries as in some other places in the world. The unwary tourist often goes on the assumption that a stone bought in a gem producing center is likely to be the real thing. You are more likely to get fair treatment from your home town jeweler whom you know, than from the hands of the unknown Oriental merchant. The fact that a stone originated in a gem producing center is no criterion of its having been mined. The natives in the gem producing centers have even been clever enough to purchase an uncut boule, shape it like a water-worn pebble and place it in a lot of mined material. The trick has worked. Remember, too, that good, genuine stones are not sold at "bargains" to tourists on the streets in an Oriental gem center; no more than one would expect to buy gold nuggets at reduced rates in a mining camp. Both have a standard value which the producer can readily obtain in the regular markets.

DICHROISM OF RUBY—Due to the peculiar splitting of corundum boules along their optic axes, it is uneconomical to cut them perpendicular to the optic axis, as the natural ruby and sapphire are generally cut. For this reason a synthetic ruby may be dichrotic through the table, while the natural gem is usually isotropic in this direction. Any stone which is claimed to be natural and which shows dichroism through the table should be carefully examined before passing judgment. A few natural rubies and sapphires are cut against the above rule for various reasons, including the presence of spots of color, flaws, and imperfections.

ALUNDUM ABRASIVE—The mineral bauxite, fused in an electric furnace of the arc type, is purified, crystallized, and becomes the familiar abrasive and polishing agent, *Alundum*.

In chemical composition the *Alundum* abrasive resembles sapphire and emery. The main characteristics of the artificial material, *Alundum*, is its toughness, hardness, and highly uniform purity and quality. It is used universally as a polishing powder in the lapidary industry.

POLISHING DATOLITE—The fine, massive, compact, specimens of datolite, showing native copper, which occur in Michigan, often present a problem to polish free of scratches. Seemingly small fragments of native copper are torn from the specimen and carried across the surface by the felt buff. Generally the copper is found along the edges of the nodular masses of datolite; hence, if the polishing is directed from the center toward the edge much of this scratching will be eliminated.

SANDING—The surface of a large specimen should be sanded in one direction, preferably along the natural pattern of the stone. This practice tends to obscure any scratches that may be left after polishing; in other words, any remaining scratches would blend in with the natural layers or lines of the agate or whatever material was being worked.

ADAPTING SANDING CLOTH—Wrinkles may appear when adapting the sanding cloth over the working face of the vertical running type of sander. In order to get a smooth adaptation of a new sanding cloth, the back (not abrasive side) side should be made slightly moist with water. This will soften the cloth and permit easier adaptation to the face of the wooden disk.

On this type of sander the cloth is generally held firmly in place by the use of a metal hoop, fitted tightly over the slightly tapered periphery of the wheel. Some workers cut a groove on the periphery of the disk, and hold the sanding cloth in position with a heavy rubber band cut from an old inner tube. In both casses the dampened cloth can be more easily adapted.

TOURMALINE CABOCHONS—Gem quality tournmaline, free of all flaws and suitable for facet cutting, is by no means common material, and generally cannot be had at a low price. Tourmaline often occurs in good quality so far as color is concerned, but all too often the crystal is found flawed, rendering it unsuitable for facet cutting. Excellent cabochons can often be worked from these crystals; small bicolored crystals where the colors meet in such a fashion as to permit finishing a stone showing equal portions of color are especially attractive. Some types of green tourmaline exhibit a fibrous effect, and this material even when flawed will frequently yield choice cabochons which present a "cat's eye" effect.

NORBIDE SPEEDS SAW—One of the chief objections to the "mud" saw is the slowness of the operation. The mud saw can be speeded up by adding a small amount of *Norbide* to the silicon carbide mixture. *Norbide*, a recently developed product of the electric furnace, is the hardest substance yet made by man on a commercial scale. It is considerably harder than silicon carbide or sapphire.

CARE OF MUD SAW—The periphery of the mud saw tends to become tapered after considerable use. Hence it is advisable to trim off an eighth or a quarter of an inch from time to time. This will also true the edge as well as remove most of the tapered portion. The trimming can be done with a sharp fractured piece of agate, holding the cutting tool on a firm steady rest against the saw edge while in motion. A sharp steel tool can also be used for the trimming. Speeds over about 300 R.P.M. (other size disks in proportion) for a twelve-inch disk will produce "flats" more rapidly.

CABOCHON BEZEL ANGLES—The beginner in cabochon cutting is often at a loss to know what angles should be made on the edge or girdle of a cabochon cut stone. This will be dependent on what purpose the stone is for. If it is merely a gem that is for display purposes only and not to be mounted, it matters little what angle is left at the "bezel" portion.

If the stone is to be mounted consideration must be given to the type of mounting to be used; otherwise the manufacturing jeweler may have difficulty in properly mounting the gem. If the stone is to be mounted in a heavy cast sterling silver mounting, then only a slight angle need be given the bezel portion. Measuring the slope from the flat base, the angle should be approximately 10 degrees. The reason only a slight angle is given for this type of mounting is the difficulty of pressing or bending a heavy mass of silver. On the other hand, if the gem is to be mounted in a gold mounting where thin strips of gold are bent by hand to form the bezel, then 20 to 30 degrees slope should be given to the side of the stone. Similar silver mounts can be given a slightly less angle.

A cabochon cut stone lacking the proper angle for the given type of mounting will tend to loosen in the setting. Generally a cabochon gem that becomes loose in the mounting can be charged to incorrect cutting or careless work in mounting, or, of course, rough use of the ring may loosen any stone in its mounting.

DRILLING HOLES

Holes are sometimes drilled through a gem stone to render it suitable for some ornamental purposes. Soft gems like malachite and turquoise can be drilled with small steel drills (dentist's type), but the harder gems require different technic. There are two types of drilling, one being done by the use of metal tubing, the other by a diamond point.

Tube Drilling

Small holes can be sunk by the use of a small section of jeweler's drill tube and silicon carbide made into a paste with oil. The new hard abrasive, *Norbide*, will work faster than silicon carbide, especially when dealing with hard substances. Small holes can be worked by the use of fine grit No. 220 to No. 320, depending upon the depth of the hole. The soft metal tubing is slightly flattened on the cutting end (cut clearance), mounted in a small drill press and fed with gentle pressure. Diamond dust, as is used for sawing, can be used as an abrasive with good effect.

Large holes can be cut with the proper gauge iron, copper, or brass tubing, keeping the cutting end slightly flat by gentle tapping with a light hammer. If the end of the tube drill fails to cut clearance for the core, it will jam and possibly lodge fast in the work. The tube must also cut a hole slightly larger than its outer diameter, to prevent wedging fast. Silicon carbide or *Norbide* or No. 120 grit is generally employed for larger holes.

By the use of two large sized tubes, one slightly smaller than the other, it is possible to cut a circular band from a flat piece of agate or jade, or any tough gem material. These are worn as a finger ring. To produce a ring of this kind, a hole of proper size (finger diameter) is first made in a flat polished piece of the material. The slightly larger tube is then used to cut the ring free of the surrounding matrix. Rings can be cut from flat sections of iron meteorites by this means. In drilling any type of hole, with either tube drills or diamond mounted points, it is advisable to clamp the work rigidly to a secure base, below the drill.

Some years ago, when the huge agatized dinosaur skeleton was mounted at the Field Museum, it was necessary to first sink large holes through the thick agatized bones, to enable articulation by passing cables through the holes. Some of the bones were of considerable thickness and fully agatized. They presented a problem in tube drilling. Ordinary iron pipes were used, with silicon carbide as the abrasive, and while the work was slow, numerous holes were drilled without any difficulty. It was found important to keep the ends of the pipe slightly blunt by hammering at intervals.

Diamond Drilling

Small holes can be sunk by the use of commercial diamond drilling points. These are mounted and ready for use. Diamond drills are generally utilized for the commercial drilling of beads where speed is a factor in production. The diamond drill is a delicate tool and requires care and skill in use.

Cutting Slots

Where a square or rectangular shaped hole is desired, this can best be accomplished by the use of small mounted silicon carbide points of the same type as are used for cameo cutting and engraving. Mounted points and small wheels are effective where the stone is not over a quarter of an inch in thickness. The slot can be worked from both sides.

CAMEO CUTTING AND CARVING

Cameo cutting and gem stone carving and engraving is an art requiring considerable skill, in order to produce artistic work on hard gems. Mounted silicon carbide wheels and points (vitrified) can be used to good effect in working the harder gems. Those softer than steel can, of course, be cut and engraved by the use of engraver's tools.

A portable, high-speed grinder, in which the points and wheels can be mounted, is very useful for cutting and engraving. The work is held in a padded vise and the grinder held by hand. Small soft metal points charged with silicon carbide, *Norbide*, or diamond are also effective in some types of cameo cutting and carving. The fine grits of silicon carbide are used to smooth the work, after the heavy cutting is completed. Small hardwood points are also used to carry both cutting and polishing abrasives to the work.

QUICK TEMPORARY POLISH

It may at times be desirable to place a temporary "shine" or polish upon a specimen, to indicate how same will appear when finished. It is customary to wet the surface of a specimen after sawing to bring out the details hidden by the rough surface, which prevents penetration of light.

Silicate of soda (water glass) has found some use in coating a cut surface to obtain a temporary polish. Ordinary varnish has also been tried. These and similar coatings have the disadvantage in that the presence of oil, kerosene or grease will not permit placing an even surface or "gloss" on the specimen. Oil and kerosene are generally used in sawing specimens, and are difficult to completely remove. Therefore an oil soluble has obvious advantages.

Dake's Varnish

A mixture of equal parts of canada balsam and xylol (*Dake's Varnish*), applied to the surface will give a very satisfactory temporary "polish." Filtered liquid canada balsam is best, but the dry balsam can be used. It is more slowly soluble in the xylol. The varnish can be applied with a soft camel's hair brush or a soft cotton swab, but best results will be obtained if applied with a spray, in the same manner as quick drying paints are applied. The surface should not be gone over more than once; otherwise streaks will appear. Dipping the specimen in the varnish is also satisfactory. It will require about two hours for the varnish to dry, but if placed in a warm place, like under a strong electric light, setting time will be lessened. The mixture can be made more concentrated for special purposes, but will require longer to set, unless placed in a warming oven.

One of the outstanding advantages of Dake's Varnish is its relatively high index of refraction, which appears to aid in bringing out a better reflection and "polished" surface. While the varnish can be applied directly on the sawed surface of the specimen, better results will be had if any deep scratches are removed by grinding on the side of the wheel or lapping on the horizontal running lap. What remaining scratches are present will be filled in and become invisible by varnishing. A fairly smooth surface, properly varnished, will defy ordinary observation as being a "doctored" specimen, and is very suitable for its intended purpose.

Large thin sections of translucent agate and similar gems are often displayed, mounted in frames, behind glass. It is often difficult and time-taking to attempt to fully polish a large flat surface of this kind. Since the specimens are protected by glass in these illuminated transparency displays, varnishing is fully as satisfactory as a regular polish, especially if the deeper scratches are removed prior to varnishing. Transparencies made of chiastolite and numerous other minerals can be varnished with good effect. Soft minerals are sometimes difficult to finish with a glossy surface, and if they are compact and not too porous, Dake's Varnish will be found effective. Smooth water-worn pebbles can also be treated in this manner.

SPHERE CUTTING

Early Cutting

With the development of simplified technic the art of sphere cutting is again coming into popularity. For the past few years sphere cutting has become very popular amongst the southern California gem cutters.

Probably the first spheres to be cut were fashioned from crystal quartz, but just when this was first done has been lost in the antiquity of time. It is well known that sphere cutting started in the Orient many centuries ago, when the Chinese cut small spheres of quartz, to be carried in the hand to cool the palms, a fashion which has continued to some extent down to the present day.

Spheres cut from quartz crystal have frequently been found in the ancient ruins, even of the Ninevites, who used the spheres as burning lenses. In the time of Pliny, surgeons used crystal spheres for cauterizing by focusing the solar rays. Orpheus recommended that a crystal sphere be used to kindle sacrificial fires, thus ensuring the favor of the gods. The flame thus being kindled was called the Fire of Vesta. At an early date seers made wide use of spheres and became known as "crystal gazers."

In recent years a good deal of sphere cutting has been carried on in old Mexico, where a colorful variety of onyx (calcite) finds wide use. The Mexican spheres evidently are cut by crude methods for they are often "one sided." A perfectly cut sphere when rolled on a level sheet of glass should follow a reasonably true course over some distance. This, of course, is only an approximate test for true cutting.

Many Materials

By the aid of the recently developed technic spheres can be readily fashioned from any of the hard gem minerals. Large gemmy garnet crystals make superb spheres, and when the specimen shows asterism the finished sphere, if over two inches in diameter, may be rather valuable. In the Museum of Natural History at Cleveland is a perfect garnet sphere, some three inches in diameter which shows excellent asterism. When this specimen is rolled slowly the flashing "star" appears repeatedly at various positions on the surface. Almost any compact gem material of sufficient size can be utilized for sphere cutting.

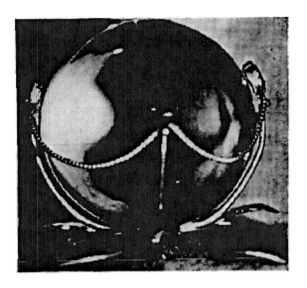

Fig. 33-A. Showing manner in which a sphere cut from a gem mineral may be mounted as an ornament. Sterling silver mounting by Mrs. A. N. Goddard, Detroit. Michigan.

The cutting of perfect spheres, contrary to general belief, is a relatively simple operation. Besides the ordinary lapidary saw, grinding wheels, and polishing buff, the only additional equipment required for sphere cutting is the special sphere cutting tool, shown in the accompanying illustration. This sphere cutting tool can be operated either in the horizontal or vertical position. Special tools of this kind are available from various lapidary supply houses.

Cube First

If the material from which the sphere is to be cut is in the form of a large mass, it is then advisable to first saw out a cube. The corners of the cube are then rounded off on the grinding wheels until an approximate sphere is formed. The rough sphere is then made circular by the special sphere cutting tool and then polished in the usual manner on the felt buff.

Many rough specimens, like some agates and nodules, are already in the spherical form. The spherical agate filling a "thunder egg" can often be utilized to good advantage.

Sphere Cutting Tool

The sphere cutting tool consists essentially of various sized iron pipes which are attached to the shaft of an electric motor. Short lengths of ordinary iron pipe are excellent. The edge of the pipe should be slightly beveled. In the illustration it will be noted that the sphere is held in position while being rotated by a second piece of iron pipe held by hand. The short lengths of iron pipe are open only at one end, and can thus be filled with the abrasive mixture. The diameter of the pipe should be slightly less than the diameter of the sphere desired.

The Craft of Jewellery Making

Fig. 33. The art of sphere cutting, showing technic used. Sphere (white) being shaped by the right hand of the operator. (Photo John Grieger, Pasadena, Calif.)

Cutting is carried out by silicon carbide of various grits. The start is made with coarse grit, and final lapping is done with fine grits. This phase of the work is the same as preparing large flat surfaces for the polishing operation.

The horizontal running lap can be easily adapted for sphere cutting. The manufacturers supply sphere cutting tools that will fit or can be fitted to the horizontal lap. It will be seen that if sphere cutting is carried out with the tool in the upright position there will be many advantages. For one thing the lower reservoir (the lower pipe) can be filled with the abrasive and water mixture, and controlled better. However, good work can be done on either the horizontal lap unit or the equipment which is attached to the motor shaft and operated in the horizontal position. Special motors are available which can be operated in the vertical position, and in this case work can be done in the vertical position.

Polishing

After all deep scratches are removed from the sphere by lapping with very fine grit silicon carbide, polishing can be done in the usual manner on the felt buff. In polishing the ball can be held by hand and rotated frequently. It will be found that the curved surface of the sphere will polish quite readily. Curved surfaces are easier to polish than large flat surfaces.

Other Methods

There are various other methods by which spheres can be cut, and a number will be referred to here.

Commercially, spheres are usually cut on what is termed a "ball mill." The material is first rough ground to an approximate sphere and then a large number are placed in the ball mill. This equipment consists of two large circular plates, one above the other. The two plates rotate in opposite directions, and abrasive grit and water is fed in at intervals. The plates can be set at any distance, and all balls will be reduced to the same diameter.

The ends of heavy glass bottles can also be utilized to hold the sphere in the same manner as the iron pipe, but the glass bottle holds no advantage and is obviously somewhat hazardous.

The principle of the use of the iron pipe is not at all new, for many centuries ago the Chinese artisan used circular sections of bamboo for sphere cutting. A section of bamboo would be split lengthwise to form a half circle trough. The trough would be charged with some abrasive like sand, garnet, or impure emery powder, and the ball worked back and forth by hand. The final shaping was carried out on the end of a complete circular section of bamboo. Since bamboo is readily available in various size sections, the method proved effective, except that the bamboo is not as uniform in diameter as modern iron pipe.

BRACELET CUTTING

Bracelets for wrist wear, cut from various semi-precious gem minerals can be worked by various lapidary methods. Perhaps the method used by the Chinese lapidarist is the most simple and effective. The Chinese often produce splendid bracelets from jade and nephrite. In selecting a gem material for bracelet cutting, the mineral should be tough, free of fractures, flaws, and cleavage. A gem mineral with a marked cleavage would be wholly unsuited to withstand wear as a bracelet, or even a finger ring.

Chinese Method

In working valuable jade, the Chinese have long since learned various methods of conserving the rough material. In hollowing out a vase, cup, or bowl, the Chinese artisan will first cut a small core to the proper depth. This core is sunk by tube drilling and then broken out with a sharp blow from a hammer. The core can of course be used for cabochons or any other suitable ornament. After the central core has been removed, a larger size tube is used to take out a hollow cylinder of the desired diameter. In order to avoid breakage when taking out the cylinder, same should be undercut, working in the space left by the original core. The undercut can be made by the use of small and thin-mounted silicon carbide wheels, on a hand grinder.

A block of rough gem material will also suffice for obtaining hollow cylinders of various sizes suitable for rings and bracelets. The same technic as described above can be applied, first removing a central core.

Tube Drilling

Tube drilling can be carried out by various methods, the power drill press being the most convenient and effective.

The Chinese use ordinary iron pipe of various diameters, and as an abrasive coarse silicon carbide grit serves as the cutting agent. Copper and brass tubing is also excellent. Diamond dust (bort) of about 120 grit is the fastest cutting agent for use in tube drilling.

In order to cut clearance the tube used for drilling should be slightly flanged on the working face. This can be done by gently tapping with a hammer. With large heavy tubes, a chisel can be used to cut notches on the working face, and a flange bent down by hammering. The notches will also serve to better hold the abrasive where the cutting is being done.

Plenty of lubricant should be applied at the working point. In drilling very small holes, light oil is mixed with the abrasive grit. For very large holes, water is a satisfactory lubricant.

Grinding Blank

After a cross section has been cut from the cylinder, it is necessary to round off the edges by grinding on the silicon carbide wheels. Holding the work by hand is unsatisfactory as it will be found difficult to keep the bracelet the same diameter throughout. If cut too deeply on one area, it will be necessary to reduce the entire surface to this same diameter.

In grinding a bracelet (or finger ring) the Chinese does not attempt to hold the work against the grinding wheel. The work is revolved against the wheel and by this method it is not at all difficult to produce a symmetrical piece of work. A method whereby the grinding wheel revolves in one direction and the work in the opposite direction would be ideal.

Polishing

After the blank cut from hollow cylinder has been properly shaped by grinding, the deep scratches should be removed by the usual sanding operation. The inner surface of the bracelet or ring can be sanded by the use of various size felt cones, and small felt wheels, using fine grit (220 or finer) silicon carbide or *Norbide*.

Final polishing is carried out on the regular felt buffs, holding the work by hand. The inner surfaces can be reached by small felt polishing wheels or cone-shaped polishing wheels. A cone-shaped polisher can be made by cementing felt or soft leather to a cone-shaped piece of wood.

Small felt buffs and cones are available from supply houses. These are intended to be used on small polishing motors and hand grinders. The regular polishing agents are indicated in this work.

Agate Rings

There are a number of gem minerals other than jade and nephrite which can be worked into bracelets and rings. Agate, free of flaws and fractures, can be utilized as this material lacks cleavage and is quite tough and tenacious. Some of the hard, tough, and compact varieties of massive garnet (grossularite) are well adapted for these ornaments. Other gem minerals will suggest themselves as being suitable.

In cutting a hollow cylinder for a bracelet due care and attention should be given to the final size to slip over the hand. The size of a finger ring can be enlarged, by grinding on the inner surface, if there is enough material to permit same without breakage. Grinding and polishing on the inner surface of ring can be done by the aid of small grinding wheels and buffs, mounted on a small mandrel, and held in a hand grinder.

Hand Grinders

Hand grinders are available from various lapidary supply houses, and are available in several styles. One type is powered by a very small electric motor contained within the unit. Another style hand grinder, has a flexible shaft which can be attached to the armature shaft of a small electric motor. Grinders of this kind find wide use in the home jewelry and lapidary shop.

LAPIDARY PENCIL

A piece of ordinary aluminum wire about three inches long and ground to a long point at one end, is very useful to the lapidarist. The pointed end is used to outline the desired final shape of the cabochon set to be cut, using it like a pencil on the flat sawed surface of the gem material. The advantage of the aluminum pencil or wire over an ordinary pencil is that it will not rub or wash off during the grinding process, and the mark can be seen when the rough gem is either wet or dry. As a matter of fact about the only way to remove the aluminum markings is to erase them with a bit of old abrasive cloth.

Another use the writer suggests for the aluminum pencil is aiding in cutting a uniform height of bezel on a cabochon cut set. While shaping out the blank to the desired size and shape for the cabochon in question, work with the back of the cabochon up and slope the cut about thirty degrees from the perpendicular. (Note: For cast mountings such as heavy sterling silver rings, it is advisable to give about fifteen degrees slope to the bezel.) We now have a blank oval, round or square, about an eighth of an inch thick and correct slope inwards around the edge. With blunt end or side of the aluminum pencil, shade the entire one-eighth inch sloping edge (more or less) of the blank with the aluminum pencil. Let us assume you wish to have a finished stone with a bezel one-sixteenth of an inch high. As you round off the top of the stone it is quite easy to see exactly how uniform you are cutting the bezel, or to put it another way, just where you have to remove material to obtain a bezel of uniform height all the way round.

LAPIDARY NIPPERS

In cutting a cabochon from a sawed slab of agate or similar semi-precious gem material it may be necessary to remove considerable waste material. In order to eliminate considerable grinding or resawing, a special type of "lapidary pliers" can be used to crush or nip off portions of the waste material. This type of pliers is in use in some commercial cutting shops, and the writer wishes to credit the suggestion given here to Oscar Smith of Smith's Agate Shop, Portland, Ore.

The jaws of the lapidary pliers should be well rounded, and the handles about six inches or more in length to give enough leverage. The rounded jaws can be about an inch in length. This type of plier can be obtained in any hardware store at a nominal cost. This tool is widely used for bending wire.

The knack of trimming a slab of agate requires a little practice, as the writer learned. Do not start operating on a valuable specimen. Select a discarded specimen for your first experience. There is, of course, a limit to the thickness of a sawed slab of agate which can be handled in this manner. Sections not over three-sixteenths of an inch can be trimmed readily. In using the tool, select an outermost corner, taking a bite about an eighth of an inch deep, and apply pressure on the plier handle. This will likely crunch off the irregular corner. When all the projecting corners have been removed by the crushing method, additional and larger fragments can be removed. Hold the agate slab firmly in one hand, apply the jaws of the tool, and with a rolling motion, a fragment about an eighth of an inch deep can be removed.

By following the above outline it will be possible to "nibble" down a fairly large section in a few minutes. After you become adept it will be possible to trim quite close to the finished size, and what is more important, save time on the grinding wheels or the resaw. Do not attempt to clip off large sections. Otherwise you may fracture the section beyond the desired point. The rounded jaws on the tool appear to be more effective in controlling the breaking, compared to ordinary jawed pliers.

TABLE OF HARDNESS AND SPECIFIC GRAVITY

Gem	S.G.	H.
Alexandrite	3.64	8.5
Almandite	3.68—4.33	7 —7.5
Amber	1.05—1.10	2 —2.5
Amazon Stone	2.54—2.57	6 —6.5
Andalusite	3.1 —3.2	7 —7.5
Apatite	3.15—3.27	4 —5
Azurite	3.77—3.83	3.5—4
Benitoite	3.64—3.65	6 —6.5
Beryl	2.63—2.91	7.5—8
Calcite	2.69—2.82	3
Cassiterite	6.8 —7.1	6 —7
Chalcedony	2.55—2.63	6 —7
Chiastolite	3.1 —3.2	7 —7.5
Chloropal	1.82—	2.5—4.5
Chrysolite	3.3 —3.5	6.5—7
Chrysoberyl	3.5 —3.84	8.5—
Chrysocolla	2.4 —2.41	2 —4
Cinnabar	8.0 —8.2	2 —2.5
Corundum	3.95—4.10	9 —
Diamond	3.15—3.52	10 —
Dumortierite	3.26—3.36	7 —
Epidote	3.06—3.5	6 —7
Fluorite	2.97—3.25	4 —
Garnet Group	3.15—4.3	6.5—7.5

Mineral	Specific Gravity	Hardness
Glass (fused)		5.5—
Grossularite	3.4 —3.6	6.5—7
Gypsum	2.3 —2.45	3 —4
Hematite	4.9 —5.3	5 —6.5
Hiddenite	3.1 —3.2	6 —7
Jadeite	3.3 —3.5	6 —7
Kunzite	3.1 —3.2	6 —7
Labradorite	2.68—2.72	5 —6
Lapis Lazuli	2.38—2.45	5 —5.5
Malachite	3.9 —4.03	3.5—4
Marcasite	4.6 —4.9	6 —6.5
Moonstone	2.5 —2.6	6 —6.5
Nephrite	2.69—3.1	6 —6.5
Obsidian		5.5—
Olivine	3.3 —3.5	6.5—7
Opal	1.9 —2.3	5.5—6.5
Peridot	3.3 —3.5	6.5—7
Phenacite	2.94—3.04	7.5—8
Pollucite	2.86—2.9	6.5—
Prehnite	2.8 —2.9	6 —6.5
Pyrite	4.95—5.16	6 —6.5
Pyrope	3.5 —3.8	7 —7.5
Quartz	2.59—2.66	7 —
Rhodochrosite	3.3 —3.76	3.5—4.5
Rhodolite	3.83—	7 —7.5
Rhodonite	3.4 —3.68	5.5—6.5
Ruby Spinel	3.52—3.71	8 —
Sapphire	3.95—4.10	9 —
Smithsonite	4.30—4.45	4.5—5
Sphalerite	3.9 —4.1	3.5—4
Spinel	3.5 —4.1	8 —
Sunstone	2.6 —2.7	6 —6.5
Thomsonite	2.19—2.4	5 —5.5
Topaz	3.4 —3.65	8 —
Tourmaline	2.9 —3.2	7 —7.5
Turquoise	2.6 —2.88	5 —6
Variscite	2.47—2.54	5 —
Vesuvianite	3.35—3.45	6.5—
Zircon	4.02—4.86	7.5—
Zoisite	3.25—3.36	6 —6.5

FACET ANGLES FOR CUT GEMS

	R. I.	*Facet*	*Angles*
Diamond	2.42	C. 35°	P. 41°
Zircon	1.92	C. 43°	P. 41°
Garnet—			
Demantoid	1.88	C. 43°	P. 40°
Spessarite	1.81	C. 43°	P. 40°
Almandine	1.79	C. 37°	P. 42°
Rhodolite	1.76	C. 37°	P. 42°
Pyrope	1.75	C. 37°	P. 42°
Hessonite	1.74	C. 37°	P. 42°
Andradite		C. 37°	P. 42°
Uvarovite		C. 37°	P. 42°

Benitoite	1.75	C. 37°	P. 42°
Chrysoberyl	1.74	C. 37°	P. 42°
Sapphire	1.76	C. 37°	P. 42°
Epidote	1.73	C. 37°	P. 42°
Spinel	1.72	C. 37°	P. 42°
Diopsite	1.68	C. 43°	P. 39°
Peridot	1.66	C. 43°	P. 39°
Phenacite	1.65	C. 43°	P. 39°
Kunzite	1.65	C. 43°	P. 39°
Tourmaline	1.62	C. 43°	P. 39°
Topaz	1.61	C. 43°	P. 39°
Beryl (all varieties)	1.57	C. 45°	P. 41°
Quartz (all varieties)	1.54	C. 45°	P. 41°

R. I.—Refractive index, *only the lower one given* where there are more than one. C—Main crown facets. P—Main pavilion facets.

In the above table only the principal facet cut gems are listed. Any gem not listed can be fitted into table by reference to its refractive index, using the lower one if there are two.

As the index of refraction of a gem stone decreases, less light is returned through the crown of the gem. Hence the lower the index of refraction of the material the less brilliancy will be possible. All stones having an R. I. of less than 1.65 will show a "well" or dull spots are some place on the crown.

COMPARATIVE HARDNESS OF STONES

The comparative hardness of gem minerals is difficult to determine (by hand testing) where there is only a slight variation. It is, however, a simple matter to separate those which vary one degree or more, like quartz from topaz. Hardness testing can be best carried out with special points made for that purpose. These are more satisfactory than attempting to place a scratch with a large blunt point. Tests made on a smooth surface can be examined with a low power glass to note any scratching. Valuable gems should be handled with care. The girdle is the place generally tested for hardness. The slight "dust" of abrasion should not be confused with an actual scratch.

Moh's scale of hardness, used in all standard texts on mineralogy, fails to properly indicate the actual difference between the various gems. On Moh's scale, sapphire is listed as nine and diamond as ten, but the latter is actually at least *ten times* harder. The tables below may serve to show the actual differences:

	Moh's	Gem Scale
Diamond	10	10,000
Norbide	9½	4.000
Sapphire	9	1,000
Topaz	8	450
Zircon	7½	350
Quartz	7	250
Agate (average)		225
Steel file		200
Opal (variable)		150
Knife blade		100
Glass (variable)		100
Malachite		0
Calcite		00

OVERLOADING DANGEROUS

Attention is called to the danger of placing too large a grinding wheel on a small diameter grinding arbor or spindle. Instances are known where a 12 inch grinding wheel has been mounted on a one-quarter or one-half inch spindle, and in operating at standard speed, would be torn off thus releasing the whirling wheel. This practice is fraught with danger.

Below are given *minimum* safe spindle sizes.

6×1 inch wheel—1/2 inch spindle.

9×1 inch wheel—3/4 inch spindle.

12×1 inch wheel—1 inch spindle.

20×1 inch wheel—1 1/2 inch spindle.

The above spindle sizes are given for a maximum operating speed of 7,000 surface feet per minute, the standard speed used being 6,000 S.F.P.M.

In the use of wet grinding wheels do not permit a wheel to stand partially immersed in water for a period of time. The water soaked portion being heavier may throw the wheel dangerously out of balance. If a wheel has been subjected to rough handling in shipping a crack may develop, hence in starting a new wheel in operation, stand to one side until after full wheel speed has developed for a few minutes. Do not remove the washers of blotting paper which are placed at the factory, this compressible material is placed there for a good reason, it is not merely ornamental. When tightening arbor (spindle) end nuts do not use more pressure than to hold the wheel firmly.

CALCULATING SPEEDS

Attention has been called to the operation of various lapidary tools at standard speeds for best performance. Grinding wheels for example, give best efficiency when operated at a *surface speed* of approximately 6,000 feet per minute, or about 1,910 revolutions per minute in the case of 12 inch wheel. It will be seen that in order to get the same efficiency, a 6 inch wheel would have to be operated at a high r.p.m. speed.

The following formula will give *surface feet per minute:* R.P.M. X, diameter X, 3.1416, divided by 12 equals, surface feet per minute.

To find the number of revolutions of wheel spindle, surface speed and diameter of wheel being known; multiply surface speed in feet per minute by 12, divide the product by 3.14, and divide again by the wheel diameter to obtain r.p.m. of wheel.

To find proper speed of countershaft (line shaft), proposed speed of grinding arbor being given: Rule—Multiply the number of revolutions per minute of the arbor by the diameter of its pulley, and divide the product by the diameter of the driving pulley on the line shaft.

To find the diameter of pulley to drive arbor, speed of line shaft being given: Rule—Multiply the number of R.P.M. of the arbor by the diameter of its pulley, and divide the product by the number of R.P.M. of line-shaft.

MOUNTING THE GEM--JEWELRY MAKING

By JOHN F. MIHELCIC, M.A., *Industrial Arts Instructor, Cooley High School, Detroit, Michigan. Lapidary Instructor, Cranbrook Institute of Science, Bloomfield Hills, Michigan.*

THE NATURAL STEP following the cutting of the gem is to place it in a setting that will enhance the inherent beauty of the stone. This takes the form of jewelry, art craft objects, or simple mountings to support the gem for display purposes. An unmounted stone lacks the sense of completeness that even a simple setting gives to it.

To aid those who wish to make suitable mountings for their stones, this article will deal first with the tools and processes involved and then the sequence of these processes in the construction of an actual ring, brooch and bracelet.

The jewelry illustrated in this section has been all hand made.

Assuming that it is necessary to have special talent to make these settings is an error that is commonly made. Acceptable jewelry and art craft settings can be readily made by performing a few simple but basic operations that anyone can learn to do in a reasonably short time with a few inexpensive tools. Talent in design is not a matter of particular concern because a scrap book of clippings from catalogues and magazines or design sheets obtainable from most any craft metal supply house will furnish an abundance of ideas for the beginner, who in time will develop a sense of the appropriateness of a proposed design.

Certain tools are necessary. These are:

Hammer	Flat Nosed Pliers
Hand Drill	Round Nosed Pliers
Snips	Flat File
Jeweler's Saw	Half Round File
Torch	Needle Files
Small vise	Pointed Tweezers
Steel Ruler	Small Drills
Center Punch	

Since the majority of these tools may be purchased in the five and ten cent store, the cost is not great. Other devices that can be readily made or improvised will be mentioned later.

The Craft of Jewellery Making

All the basic processes of jewelry making can be executed with this equipment. These processes are:

Sawing	Soldering
Annealing	Cleaning
Forming	Polishing
Embossing	Coloring
Carving	Setting
Wire Twisting	

The amount of space needed for the jewelry workshop need not exceed that of a kitchen table, as a matter of fact, a kitchen table and gas stove have been instrumental in the creation of some very fine jewelry.

Fig. 40. John Mihelcic at work at his home jewelry making bench.

The first thing to be considered in the construction of a project is the design, which should conform to the use to which the article will be put, to the age and sex and personal characteristics of the person for which it is designed and to the shape and the quality of the stone being used. Since the three most common forms of jewelry for wear are brooches, rings and bracelets, their design must take into consideration these factors.

A ring must be circular to fit over the finger; the inside should be smooth with rounded edges; the portions that come in contact with other fingers should have no sharp edges or be so heavy as to interfere with the bending of the finger; and the stone and ornament should not be so large as to interfere with the freedom of movement or, in some cases, with the wearing of gloves. A man's ring should be quite heavy and bold in outline, while a lady's ring is more delicate even though the stone may be large. A child's ring is usually very simple in character. Costume jewelry requires greater size and color. Since irregularly shaped stones are rarely used in rings, there is no problem except that of the appropriateness of the setting of facet or cabochon cut stones.

The brooch is primarily designed to hold fabrics together and must be constructed to avoid projections that would catch and tear. It must be made strong enough to hold its shape. The principles that govern the design of the brooch are the same governing the pendant with the exception as to size, the pendant being worn at the waist line is usually large. Irregularly shaped stones often furnish the motive of the ornament.

Bracelets have been designed in many different sizes which are determined by the occasion as well as the dress. Again the inside should be smooth with rounded edges, and sharp projections that cut or tear are to be avoided. The stones are regular in outline.

Once these factors have determined the limitations of the article, the chief problem of design remaining is the ornament. The illustrations in this chapter will offer suggestions. Other sources of ideas are the jewelry catalogues and magazines, museums, and, of course, actual jewelry. It is possible to purchase shanks of rings to which may be added a few beads of silver or leaf forms. A never ending source of patterns to follow may be found in cross sections of seed pods and buds.

Transferring the design to the metal can be done in several ways. The first procedure is the tracing of the original design upon tracing paper, which should be carefully done to avoid distortion. Remove grease from the metal by any convenient method. Then glue the tracing on with diluted liquid glue. Make sure that all points are in contact. A less desirable method is to transfer the design directly to the metal with carbon paper. An improvement would be to paint the metal with Chinese white water color and then using carbon paper. The purple dye layout fluid used in machine shop practice makes a good background for scribed lines and is particularly good for detailed designs. It is also possible to follow the outlines with fine punch marks placed closely together. Avoid heavy blows in this method.

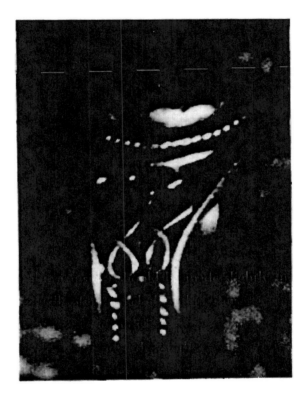

Fig. 38. Ornate ring from the workshop of John Mihelcic.

The next procedure requires drilling and sawing. Make punch marks in the open spots of the design to provide a start for the drill, the size of which may range from one sixteenth to an eighth of an inch. It is well to have about half a dozen drills of assorted sizes. The drilled holes are made to admit the saw blade. The saw, most frequently used, has a depth of about five inches and the blades best suited for the beginner are No. 1 and No. 6, the first being the finer. While there are finer and coarser blades it is well to avoid them at first.

Follow this procedure in sawing.
1. Open the saw frame a little more than is necessary to admit the blade. This is done to provide sufficient tension on the blade.
2. Insert the saw blade, teeth facing the handle, and secure it at one end.
3. Put the saw blade through one of the holes in the center of the design.
4. Placing the fore part of the saw against the bench, press forward on the saw until the blade enters the locking device, and tighten the screw.
5. Place the metal on a projecting notched board with the saw handle down, since the sawing is done on the down stroke.
6. Proceed to saw, keeping the saw vertical and taking short strokes. Slow up for curves and, naturally, keep cutting in the waste material.
7. Turn the metal and not the saw.
8. Keep close to the board when sawing to avoid excessive

pinching of the blade. A little wax on the blade will help.
9. Saw away from you, keeping a firm pressure on the metal.

After the sawing is completed, file smooth all ragged edges with a needle file. These files are used by jewelers and may be obtained in many different sizes and shapes. A surprising amount of work can be done with just a few shapes such as half round, knife and rat tail. Much of the filing can be done with the work supported on the notched board. For heavier work use the vise. Be sure that the jaws of the vise are smooth, otherwise unsightly mars may result. A good precaution is to make wooden vise linings that can be quickly inserted and held in place with simple clips. Always place pressure on the forward strokes and none on the return. The burr that results can be scraped or sanded away. Clogging of files can be delayed by rubbing chalk into the teeth, but once they are clogged a file card or wire brush will be necessary to clean them.

The one process that most people find tricky is hard soldering. For that reason time spent in practice soldering of strips of copper and brass is well worth while. Experiment with pieces of different shapes until the proper positions for effective soldering are determined.

Soft solder, an alloy of lead and tin, is not used to any great extent in jewelry and is not practical where strength is required. Hard solders are used for gold and silver articles.

Silver solder is sold in three types, easy, medium, and hard flowing. Where it is necessary to make several joints the first are soldered with hard flowing solder and are then covered with ochre or loam while the other joints are soldered with medium or easy flowing solder. Gold articles are soldered with gold solder of three or more karats lower than that of the article. The solders are generally sold in 28 gauge sheets. Do not use the bar or wire solder used for steel articles.

If these directions are followed the difficulties of soldering will be reduced to a minimum.

1. Boil the article to be soldered in a copper pan containing a solution called a *pickle*, which is made of one part sulphuric acid to twenty parts of water. (A copper pan can be easily made if necessary by doming out a piece of roofing copper.) This pickle may be used hot or cold, the cold taking a little longer to clean the metal.
2. Remove the article from the pickle with a copper hook or copper tongs. Do not use iron tongs or the article will become copper plated.
3. Thorouhly wash with water, and dry.
4. File or scrape the joints to the bright metal.
5. Borax sold in grocery stores is quite satisfactory, but a better quality can be purchased in cake form from jewelry supply houses. The borax is mixed with water to form a flux. Satisfactory solutions can be purchased.
6. Scrape the sheet of solder on both sides and cut enough

sixteenth inch squares with the snips.
7. Place these squares of solder into a saucer containing the flux.
8. Bind the pieces to be soldered with 28 gauge iron wire, if necessary, and place them on an asbestos block. A prepared charcoal block gives a better surface because of the heat radiated, but asbestos is more easily obtained and lasts longer.
9. Apply the flux to the joints with a small brush.
10. Use the brush or tweezers to place the solder squares in contact with the joint.
11. Light blow torch. The source of heat may be natural gas or some prepared gas as Prestolite, or it may be an alcohol torch. Removing the burners from a kitchen gas stove is easy, and the torch can be readily connected.

The Craft of Jewellery Making

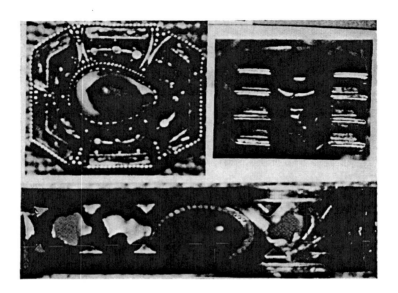

Fig. 37. Brooch (upper left) with mounted stone. Ensignia cast ring (upper right). Bracelet (lower) with mounted cabochon. Jewelry craft by John F. Mihelcic.

12. Regulate torch until you have a blue flame and apply heat to the heaviest piece first, working your flame around until the moisture in the flux has evaporated. The heat may then be applied to the joint until the metal reaches the melting point of the solder. Keep an even heat at all points to be soldered so that the solder may flow evenly.
13. More solder can be applied during this operation, but the torch should not be taken away.
14. If upon examination it is found that all the solder did not melt, apply more flux and repeat the heating.

When the article has been satisfactorily soldered it should be placed in the pickle again and cleaned. Rinse and dry. All excess solder should be scraped or filed off, and the surface smoothed with fine pumice, or sheets of wet or dry sand paper of 320, 400, and 600 grit, which are very effective means for removing scratches when they are backed up with wooden blocks. Scrapers can be made by grinding old files.

The setting of the stone is not nearly as difficult as it may appear. The bezel or box in which the stone is set can be made of a specially rolled bezel strip that is being sold by many concerns. While fine silver is nice to have for the construction of the bezel, it is not necessary, a strip of 28 gauge sterling silver will do.

To construct the bezel, get the length of the strip needed by bending it around the stone and scribing a line at the point of juncture. Cut the strip slightly large to allow for squaring up the joint. Proceed to solder it and then test for the fit. If it is too small, slip it on a steel mandrel and tap lightly with a steel hammer. This will stretch the metal. If it is too large, it will need to be cut and resoldered.

To improve the appearance of the stone and assuring a level base when the bezel itself is soldered to a rounded surface, a second band of metal, narrower than the first band is soldered inside of it. The bezel should have enough metal above the bearing to hold the stone securely. Cotter pins make handy clamps.

In case of square and oblong stones, the procedure is similar except that it is necessary to cut notches at the top of the corners to allow for the folding of the retaining metal.

In the case of bezels for faceted stones, it is necessary to build a frustrum of a hollowed cone or pyramid first and then sawing or filing the required prongs. The appearance of the bezel may be improved by cutting, and drilling.

Binding wire should be used to hold the bezel in the proper position while soldering. Avoid excessive force in binding, because unsightly cuts can be caused by the binding wire pressing against the soft expanded silver. The solder should be placed inside of the bezel.

The use of wire for decorations seems to know no limit. It may be twisted, coiled, flattened, half round, beaded or any number of forms. Jewelry supply houses furnish it in many gauges and designs, if one does not want to make up his own. If much work is to be done in forming wire, it is necessary to have draw plates to reduce and shape the wire. Twisting wire is most easily done by securing one end of a doubled wire in a vise while the other is held in the chuck of a hand drill. A very fine twist would require annealing the wire by heating it while coiled and then dropping it into a pickle.

Wire is useful in making beads of a uniform size. This is done by coiling the wire around a pencil or any small cylinder and then cutting through the coil. The rings can be converted into beads by placing them on the soldering block and melting them. Another use for the rings thus formed is found in chain making.

Duplicate decorative designs may be made of sheet metal by placing the metal over a lead block and then stamping them out with a punch bearing the desired design. While tool steel rods are the best, temporary punches can be easily shaped out of mild steel rods.

After all the ornamental work has been done, and the stone rechecked for fit (excess solder may reduce the space in the bezel) the very important operation of finish must be considered. First the article must be brought to a high polish. The natural sequence is tripoli then rouge. These come in the form of buffing bricks and are applied to cloth wheels or in some cases to chamois pads. Deeply recessed spots are reached with bristle wheels. Carefully remove all traces of the polishing compounds with ammonia.

The coloring operation enhances the appearance of the decoration and at the same times removes the harshness of shiny metal. Silver is colored with a solution of potassium sulphide, commonly called liver of sulphur. The proportions are an ounce of liver of sulphur to a quart of boiling water. While it is still hot, it is brushed onto the article until a desirable color is obtained. Test the depth of the color with some fine pumice, if it rubs off too easily, apply more of the solution. Polish the high spots lightly.

The final operation is the setting of the stone. A good

bearing for the article is obtained by melting flake shellac on a block of wood held in a vise. While the shellac is still soft, press the object onto it until all parts are supported and permit the shellac to harden. Place the stone into the bezel and with a polished rod called a burnisher, push the metal over the edge of the stone until it is in contact at all points. The best method is to start pressing at points opposite each other and then continuing until all parts of the bezel edge touch the stone. Smooth the edge by rubbing with the burnisher. Remove the article from its bed by carefully heating the shellac and avoid heating the stone. The adhering shellac can be dissolved in a bath of alcohol.

To furnish the beginner with the method in the construction of specific projects, the illustrated ring, brooch and bracelet have been selected as typical subjects. If they are made by following the instructions given here, other methods will suggest themselves, for there is no one specific way of making jewelry.

The Ring

A sketch of this ring is first made to make sure that errors of design would be on paper. Then a band of paper is wrapped about the knuckle of the finger on which it is to be worn and the correct size is marked off. Fold another piece of paper into four parts; draw the outline of one quarter of the ring on one section and transfer it to the other sections by means of carbon paper. Transfer this outline to a piece of 20 gauge metal by any of the methods described. Place the metal on the notched board and proceed to drill and saw out the blank. It may be advisable to cut out a hole directly under the position of the stone to facilitate any removal of the stone from the bezel. Smooth all the rough edges. Make a groove in a block of wood about three quarters of an inch wide and three eighths of an inch deep, and place the ring blank centered over it. Bend the blank by driving it into the groove, using a wood or metal rod and a hammer. Complete the circle by tapping the metal around the ring mandrel or steel rod. Hold the ends together with binding wire; place the ring on the soldering block with the joint upward; apply the flux and silver solder at the joint; and solder in the manner described.

Next, make the bezel of either 24 gauge silver or specially formed bezel strips which are very convenient. Bend the strip to fit the stone and cut it to the right length. Bind it with wire and solder. Repeat for the bearing ring which is set inside the bezel. File the lower edge of the bezel with a half round file until it makes a perfect fit with the ring blank. Bind the bezel to the ring and solder it in place.

Double up and twist some 24 gauge wire. Place it on a soldering block and melt a few small pieces of silver solder along it to hold the strands together. Make enough for the bezel and shank.

Take some 20 gauge silver wire and bend it around the ring

blank along both edges as is shown in the illustration. File flat the part that comes in contact with the ring.

Now assemble a ring of twisted wire that just encircles the bezel, and a plain 24 gauge wire that snugly encircles both, and the plain wire forms just made. Hold all in place with binding or small cotter pins, apply flux and solder one side at a time using solder sparingly.

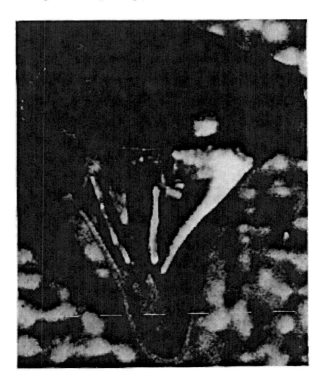

Fig. 39. Cast type of ring, by John Milhelcic.

Cut pieces of twisted wire to fit just inside the edging along the shank, make the plain wire decoration conform to the space between the bezel and the shank. Hold in place as before and solder, again one side at a time.

Remove the borax glaze by boiling it in pickle and then clean off excess solder by filing, scraping or sanding. The bezel is filed in scallops and the bearing is checked for excess solder. Proceed to polish and color as described.

In case of a ring, it is well to mount it on an arbor when setting the stone to avoid distortion. When the stone is set, all that remains is the final touching up and polishing of the highlights, followed by an ammonia bath to remove all traces of the polish.

The Brooch

Brooches vary considerably in size and amount of decoration, and while the illustrated brooch proved very popular, it may be that a simpler design would be better for the beginner. However, the sequence of operations will remain the same.

It will be noticed that the stone furnishes the basis of the design and the metal supports the shape suggested by the stone. Larger sized stones may be used, and several ideas are suggested in the illustrations.

Brooches are made up of three parts; the fastening device, a pin or clip, the base, and the ornament. It is practical to purchase the fastening devices.

In this case, the natural beauty of the stone, a Lake Superior agate, would have been destroyed by further cutting or changing of its shape. However, for a brooch, it is ideal.

To duplicate the brooch construction here, a sketch should be made before any work is started. Make this full size. Transfer the design of the base by any of the methods described, after the 20 gauge silver piece has been annealed by heating to a red heat and then dropping it into the pickle. Dome the center by tapping it with the ball end of a ball peen hammer until it is raised about three sixteenths of an inch. Saw out the open parts with a jeweler's saw and file all rough edges. Bring the outline to a true shape by sawing and filing.

Form the bezel in the manner described. Make up enough twisted wire to encircle the brooch twice as well as a ring about the bezel. Secure the bezel and twisted wire in position with binding wire and proceed to solder.

Prepare some beads from scrap metal or wire and sort them by size. It will be found that the portion of the bead that is in contact with the soldering block will be flat, thus providing a good base. Now fit the five pair of plain 20 gauge wires that extend from the edge to the bezel. By making a snug fit they will stay in place without binding. Place the beads in the position shown in the illustration, cover with flux after making sure that the points of contact are bright and clean. Use small pieces of solder. Move your torch from one section to another until all are soldered.

Fit plain 18 gauge wires in the openings to form a tight fit, and then proceed to solder 20 gauge wire "hooks" onto each. These are then carefully fitted so that they reach across the opening with a snug fit. Make all these inserts at one time and bind them into place with binding wire. Use easy flowing solder and cover up previously soldered joints with yellow ochre or loam. Move the torch from one section to the next until these pieces are all soldered into place.

Punch out designs over a lead block, place a small piece of easy flowing solder under each, and using the minimum of flux, solder them without binding. This last portion requires careful manipulation of the torch and really should not be attempted yet, if any difficulty is encountered in the soldering process.

Turn the brooch over and locate the position for the pin. Remove the pin from the hinge and solder the hinge into place, using the minimum of heat. Replace the pin and locate the position of the catch. Solder this into place making sure that the pin is out of the way and that no solder be permitted to enter either the hinge or the catch.

Clean, polish and color. Then place onto a bed on orange shellac that is softened by heat until the brooch is held firmly in place and there is support under the bezel. Set the stone in the manner described. Carefully soften the shellac without injuring the stone. Remove all excess shellac in a bath of alcohol.

The Bracelet

This article requires very little information, and in some senses is the easiest of all to make. The only precaution to take is avoiding projections that tend to snag.

The blank of silver in this case was 20 gauge silver, seven eighths of an inch wide and five and a half inches long. The design was put on tracing paper which in turn was glued to the blank with diluted glue. The blank was then placed on the notched board or bench pin as it is sometimes called. Here the openings were sawed out and then carefully filed to remove any possible roughness. With the aid of a straight edge and a scriber mark off the lines along the edges. Repeat until they are the desired depth.

A wooden mallet or a soft surfaced hammer was used to bend the ends of the bracelet over a rounded surface. It should be tested for fit, but frequently a good guess as to size is sufficient because this type of bracelet can be adjusted easily by the wearer.

Construct the bezel as before and add the twisted wire ring about it.

If the bracelet is held upright with wire supports no particular difficulty will be encountered in soldering the bezel into place.

Follow the methods for cleaning, polishing and coloring.

For those who wish to delve more deeply into the subject, the following bibliography is offered.

JEWELRY MAKING AND DESIGN, by Augustus F. Rose and Antonio Cirino.
EDUCATIONAL METAL CRAFT, by P. Wylie Davidson.
METALCRAFT AND JEWELRY, by Emil F. Kronquist.
HAND-WROUGHT JEWELRY, by Sorensen-Vaughan.

READY MADE MOUNTINGS

Many amateur gem cutters wish to mount into rings the cabochon stones produced in their home shops. Ready made mountings for cabochon stones are available from supply houses. Ready made mountings are made in standard sizes (millimeter sizes), and in sterling silver, gold, and gold filled. The sizes and styles include oval, rectangular, and square shapes. The silver mountings are inexpensive and attractive, and can also be had with inlays of ornate gold.

It is a simple matter to cut a cabochon to shape and size for a ready made ring mounting. No soldering is necessary; a wooden clamp to hold the ring mounting and a bezel closing tool are all the equipment needed. The ready made mountings are available from supply houses in numerous styles, sizes, and shapes for both ladies' and men's rings. Similar ready made brooches are also available.

For those who desire to assemble and size mountings,

shanks, bezel strips, solder, and other findings, in both gold and silver, can be obtained from supply houses. Only a few inexpensive tools are needed for the more simple mountings. A foot bellows and soldering torch connected to natural or artificial gas will suffice for soldering gold or silver.

METALS USED FOR GEM MOUNTINGS

The metals used most extensively for gem mountings are gold, platinum, and silver. The more expensive gems are mounted in gold or platinum. These metals do not tarnish as does silver.

Gold.—Pure gold is much too soft (hardness 2 1/2 to 3) and malleable to be durable as a mounting. It must therefore be alloyed with other metals to increase its hardness. The gold content of these alloys, that is, their *fineness* or *purity*, is indicated by the use of the term *carat* or *karat* (p. 135), which means one twenty-fourth part. Thus 18-karat gold, usually stamped 18K, consists of 18 parts of gold and 6 parts of alloyed metals. The use of 6 parts of alloyed metals with 18 parts of gold increases the hardness sufficiently so that the alloyed gold has satisfactory wearing qualities. In order to reduce the cost, alloys of lesser gold content are also used. Although custom and practice permit 10K alloys, when so stamped, to be sold as gold, an alloy with less than 50 per cent gold (12K) is not properly designated as gold.

The term *fine gold* is frequently used to indicate pure gold (24K). Fineness may also be expressed in terms of parts per thousand; thus, 750 fine means that the alloy consists of 750 parts of gold and 250 parts of alloys, in other words, 18K.

When gold is alloyed with different metals, very noticeable changes in color result. These alloys are known as *yellow, white*, and *green* gold.

Yellow Gold.—When equal parts of silver and copper are alloyed with the requisite amounts of the precious metal, yellow gold is obtained. Thus, 18K yellow gold consists of 18 parts of Gold, 3 parts of silver, and 3 parts of copper by weight. Lighter shades of yellow may be obtained by increasing the amount of silver and decreasing the copper proportionately without changing the amount of gold used. Darker shades result when the copper content is increased and the silver decreased. In yellow gold of lower karat rating, zinc is sometimes added to improve its working qualities.

White Gold.—This alloy contains nickel, copper, and zinc. Thus, 18-karat white gold commonly consists of 75 per cent pure gold, 17 per cent nickel, 2.5 per cent copper, and 5.5 per cent zinc. This gold alloy is somewhat harder and therefore more durable than those of platinum and iridium and is used as a substitute for them. White gold may also be made by alloying 15 per cent or more of palladium with gold.

Green Gold.—The alloy consisting of 75 per cent gold, 22.5 per cent silver, 1.5 per cent nickel, and 1.0 per cent copper is called green gold. Formerly, it was used quite extensively for mounting purposes.

Gold metal designated as *rolled gold, gold filled, gold plate*, and *gilt* is frequently used for the mounting of gems.

Rolled Gold.—This consists of an outer layer of gold alloy (for example, 18K) and an inner layer of base metal. The layer of gold alloy is "sweated" under proper conditions to a bar of base metal. This composite bar is then rolled into thinner bars or sheets or drawn into wire. In these processes the proportion of gold alloy to base metal is preserved unchanged. Articles made from rolled gold should be properly designated with respect to the proportion of gold alloy to base metal, by weight, as well as to the fineness of the gold alloy. Thus, 1/10 18K, means that the ratio of gold alloy to base metal is 1 to 9 and that the fineness of the gold surface is 18 karat. The term *gold filled* is often used for this type of composite metal. This term does not, however, adequately describe its character. Such metal is well described as *rolled gold plate* and is thus distinguished from metals that are gold-plated electrolytically and are designated as *gold plate*. Metals with very thin coatings of gold, produced either electrolytically or by merely dipping or "washing" the metal in a solution of gold, are commonly called *gilt* or *washed gold.*

Platinum.—Like gold, pure platinum is soft and flexible. For use in jewelry it must be alloyed with a metal which will impart the necessary hardness and rigidity. Alloyed with 10 per cent iridium, the hardness of platinum, which is about 4 to 4 1/2, is materially increased. This iridium-platinum alloy is admirably adapted for the mounting of gems. Because iridium is more expensive than platinum, an alloy with but

5 per cent iridium is often used, but this cheaper alloy is considerably inferior in hardness. The other members of the platinum group of metals, osmium, ruthenium, rhodium, and palladium, can also be alloyed with platinum. Because of their superior hardness, iridium (6 1/2) and osmium (7) are used for the points of fountain pens. Rhodium alloys find application in the chemical industry.

Silver.—This metal is not used extensively for the mounting of gems for personal adornment. Large quantities, however, are used for articles sold by jewelers and for coinage purposes. Silver, like gold and platinum, is soft (2 1/2) and must be alloyed with a hardening metal, usually copper. The important alloys are *sterling silver* and *coin silver*.

Sterling Silver.—This alloy consists of 92.5 per cent pure silver and 7.5 per cent copper. It is commonly stamped "sterling."

Coin Silver.—The silver coins of the United States contain 90 per cent silver and 10 per cent copper. The content of silver coins is not uniform for all countries; for example, Great Britain uses sterling silver.

Other Metals.—During the war, restrictions were placed on the use of gold and platinum for jewelry purposes. Consequently, there was a greatly increased use of palladium for gem mountings. Palladium is much like platinum in appearance and hardness, and it is many times rarer than gold. Palladium alloys can be readily distinguished from those of platinum and gold, for palladium is much lighter in weight. The specific gravities of the three metals are platinum 21.4, gold 19.3, and palladium 12. After the close of the war, platinum mountings were again in popular demand.

AFTER primitive man had satisfied his greatest need—food—his thoughts undoubtedly turned to ornaments, which were claws and tusks of wild animals hung about his body. They were worn for two reasons: first, to show his prowess as a hunter, and, second, because of a superstitition that they would help him in combat against wild animals.

Ornaments for personal adornment were in use long before clothing was used. They were of the type that adorned the neck, ankles, arms, and fingers. As man began to wear clothing, other types of ornaments such as pins and brooches came into use. Among primitive people the men wore most of the ornaments, as is true today among uncivilized people.

A study of history down through the ages shows that as man became acquainted with new materials and their uses, he immediately used that knowledge in fashioning his ornaments. The greatest advance was made in the Bronze Age, when man fashioned bronze into various forms of ornaments by hammering, riveting, and casting.

Much of the jewelry worn in ancient eras was cumber-some. At one period some of the finger rings worn weighed as much as half a pound each, and the finger band was so wide that the joint of the finger was covered. When rings with stone sets first made their appearance in Rome, it was the custom to wear them on every finger and to change them with the seasons.

Fig. 34.—Necklace made by Edward Bush, seventeen-year-old-high school student.

TOOLS AND EQUIPMENT

The making of simple handmade jewelry does not require experience, nor does it require an expensive layout of tools and equipment.

Many of the tools needed for jewelry work are already at hand in almost any school or craftsman's shop. The essential tools are as follows:

Blowpipe. For hard-soldering a gas blowpipe or an alcohol, gasoline, or acetylene torch is needed. If gas is available, the blowpipe is preferable.

Soldering Block. Charcoal blocks are very good to place the work upon while it is being heated to hard-solder, as the charcoal glows and reflects the heat back onto the work. An asbestos block or a magnesium block with asbestos fiber is also good.

Pliers. An assortment of pliers for various types of work usually includes a flat-nosed plier with squared ends; a round-nosed plier, the jaws of which are wholly round and taper toward the tips; a half-round-nosed plier, one jaw of which is rounded and has a convex surface while the other is flat; a chain-nosed plier, the jaws of which have flat gripping surfaces that taper to narrow tips and the backs or outer surfaces of which are rounded. Most jewelry pliers are available in 4, 4 1/2, and 5 inches lengths. End- or side-cutting nippers are almost indispensible in jewelry work.

Pliers with smooth gripping surfaces are ideal for jewelry

work, because the serrations on most pliers, although ideal for gripping a surface, will mar silver and gold. If your pliers have serrations, grind the jaws smooth and polish the surfaces with an abrasive cloth or an oilstone. It is much easier to prevent scratches on silver and gold than it is to remove the scratches after they have been made.

Saw Frame. A frame is essential for holding the jewelers' saw blades used in sawing out designs, ring blanks, etc. These frames are available in 3, 4, 5, 6, and 8 inches depths. For all-round jewelry work the 5-inches-depth frame is preferable.

Jewelers' Saw Blades. Jewelers' saw blades are 5 inches in length and are available in a number of sizes ranging from No. 8/0, the smallest, to No. 14, the largest. No. 0 or No. 2/0 blades are good for sawing 18- and 20-gauge silver. Smaller sized blades are used on thinner metal.

Ring Mandrel. A ring mandrel (Fig. 35) is used in shaping and forming a ring blank. It is usually 12 or 14 inches in length, made of steel, circular in the cross section, and it tapers from 1 inch to 1/2 inch. A ring mandrel can easily be made on a machinists' lathe. It may be purchased plain or graduated to the scale of U. S. standard ring sizes. The plain mandrel is available either hardened or not hardened. The graduated mandrel is hardened.

Fig. 35.—Ring mandrel. Used in shaping a ring blank.

Ring Gauge. A ring gauge or ring stick (Fig. 36) is used for measuring the size of rings. It cannot be used to form a ring as, being hollow, it is usually made of thin material.

Another use of the ring gauge is the determination, by the use of the scale on the gauge, of the length of a ring blank before the blank is formed into a ring. Measure from the metal tip of the gauge to the desired ring size on the small scale. The scale is shown in Fig. 39.

Fig. 36.—Ring stick. Used in finding the size of a ring.

Burnisher. Oval steel burnishers with wood handles (Fig. 37) are used for turning, or burnishing, the top edge of the bezel over the stone to hold the stone in place. A burnisher may have either a straight or a curved blade.

Fig. 37.—Burnisher.

Ring Sizes. Ring sizes (Fig. 38) consist of a number of metal rings, each marked with a standard ring size, which are slipped on the finger to determine the size of the ring desired.

The finger size can also be determined by the use of the scale shown in Fig. 39. Cut a strip of paper that will go just around the largest part of the finger and then measure this strip of paper on the scale to determine the ring size.

Fig. 38.—Ring sizes are used in measuring the finger to determine the size of ring.

Fig. 39.—Actual size scale showing the length of blank required for various ring sizes.

Hammers and Mallets. Hammers that are useful in jewelry making include small ball peen hammers and the French chasing hammer. Small-sized wooden mallets are useful in many ways. Other mallets that are often found convenient are those that are made of rawhide or that have rawhide or fiber tips.

Files. Files are a necessity in jewelry making. An assortment should include at least one flat, one half-round smooth file of 5- or 6-inch length, and a number of needle files.

Needle files are approximately 5 1/2 inches in length, have a round handle, and are available in a number of shapes, such as round, half-round, flat, flat-tapered, square, knife, three-square, crossing, and slitting. The half-round, flat-tapered, knife, round, and square files are used more than any of the others. Needle files may be purchased singly or in a set that consists of a dozen assorted files.

Wire Gauge. The gauge size, thickness, of wire and sheet metal can be determined by using a wire and sheet-metal gauge (Fig. 40). The gauge sizes are stamped on one side and their decimal equivalents on the reverse side. In gauging the size of wire, slip the wire into the slot, just as you do with sheet metal.

Fig. 40.—Wire and sheet-metal gauge. (***Courtesy of Brown & Sharpe Manufacturing Company, Providence, R.I.***)

Drills. Twist drills, of the smaller sizes, are needed in making many pieces of jewelry. These drills may be purchased in fractions of an 3/32 inch, such as inch, or by number.

Numbered drills run from 1 to 80.

No. by gauge	Decimals of 1 inch	No. by gauge	Decimals of 1 inch	No. by gauge	Decimals of 1 inch	No. by gauge	Decimals of 1 inch
1	0.2280	21	0.1590	41	0.0960	61	0.0390
2	0.2210	22	0.1570	42	0.0935	62	0.0380
3	0.2130	23	0.1540	43	0.0890	63	0.0370
4	0.2090	24	0.1520	44	0.0860	64	0.0360
5	0.2055	25	0.1495	45	0.0820	65	0.0350
6	0.2040	26	0.1470	46	0.0810	66	0.0330
7	0.2010	27	0.1440	47	0.0785	67	0.0320
8	0.1990	28	0.1405	48	0.0760	68	0.0310
9	0.1960	29	0.1360	49	0.0730	69	0.0292
10	0.1935	30	0.1285	50	0.0700	70	0.0280
11	0.1910	31	0.1200	51	0.0670	71	0.0260
12	0.1890	32	0.1160	52	0.0635	72	0.0250
13	0.1850	33	0.1130	53	0.0595	73	0.0240
14	0.1820	34	0.1110	54	0.0550	74	0.0225
15	0.1800	35	0.1100	55	0.0520	75	0.0210
16	0.1770	36	0.1065	56	0.0465	76	0.0200
17	0.1730	37	0.1040	57	0.0430	77	0.0180
18	0.1695	38	0.1015	58	0.0420	78	0.0160
19	0.1660	39	0.0995	59	0.0410	79	0.0145
20	0.1610	40	0.0980	60	0.0400	80	0.0135

Fig. 41.—Numbered drill sizes.

Before starting to drill a hole in a piece of metal, make a prick mark with a center punch.

Other Tools. Additional tools that are useful and oftentimes essential include tweezers, a pin or hand vise, snips, a bench vise, dapping punches, chasing tools, engraving tools, draw plates, and draw tongs.

The silver jewelry shown in the accompanying illustrations was made by students, unless otherwise credited. A minimum number of tools were used. Some of the pieces were made at home by students using a few files, a hammer, a ring mandrel, and in some instances an alcohol blowtorch.

Although all the articles illustrated were made from sterling silver, gold could be used equally well. Gold is worked in almost the same manner as silver, except that gold solder must be used. Otherwise a white streak will appear on all joints.

SILVER SOLDERING

In jewelry making one must master the art of hard-soldering, known also as silver soldering, for it is by this means that joints are made.

Silver solder is an alloy of silver, copper, and zinc and melts at a high temperature, from 1300 to 1600°F. Because of the neat, strong joint it makes, silver soldering is used widely in commercial work of all kinds. Owing to its strength and resistance to vibration, the United States government makes mandatory the use of silver solder for fuel oil lines and many other connections in airplanes for government use.

Caution: Do not attempt to silver solder a joint if soft solder is present on the piece being soldered. Under the high temperature required for silver soldering, the soft solder will react with the silver to cause serious damage, ruining the work.

Methods of Heating. Melting hard (silver) solder requires intense heat. The heat in commercial manufacturing plants is usually applied with oxyacetylene, oxygen and gas, or air and gas torches.

In school shops or the home workshop, where illuminating (artificial) gas is available and where most of the soldering is on small pieces, such as rings and bracelets, the mouth blowpipe is excellent (Fig. 42). Unfortunately, however, this type of torch is not suitable for most natural gases, although it does work well with some.

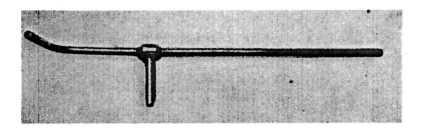

Fig. 42.—Blowpipe suitable for hard soldering where artificial gas is available.

This type of blowpipe is connected with a rubber tube to a gas outlet. The torch is lighted, the amount of gas flowing through the blowpipe is regulated, and then by blowing through the end of the blowpipe one obtains a high heat. Figure 49 shows the method of using the blowpipe. This type of torch may be used at home by connecting it to any convenient gas outlet or to the kitchen stove, after the removal of a burner which usually slips over the gas stopcock. For this torch individual mouth-pieces are available.

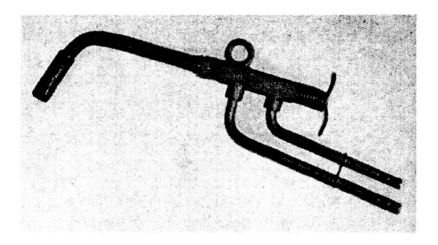

Fig. 43.—The Automaton torch which requires stream of air supplied by foot bellows or blower.

The Craft of Jewellery Making

Fig. 44.—Rotary air blower.

In using a blowpipe of this type, the user must learn to blow and breathe at the same time, for a continuous stream of air through the blowpipe is necessary. This may seem difficult to obtain at first, but by the use of the tongue as a valve to the throat it will, after a little practice, become quite easy. Stopping to breathe allows the metal to cool and often results in failure to solder the joint.

The Craft of Jewellery Making

Fig. 45.—Blower made from vacuum-cleaner motor.

The Automaton, Hi-Heat, or a similar torch may be used if much soldering is to be done, especially on large pieces. The air for such torches is supplied by a foot bellows or a blower. The size of the flame can be changed instantly by a slight movement of the fingers controlling the gas valve.

Instead of the foot bellows, a rotary air blower powered with an electric motor may be used. These blowers are available in various sizes. Many craftsmen improvise a blower from an old vacuum cleaner and motor or from a spray-gun outfit such as is used for painting. In a motor-driven blower the amount of air is regulated by installing a foot rheostat to regulate the speed of the motor.

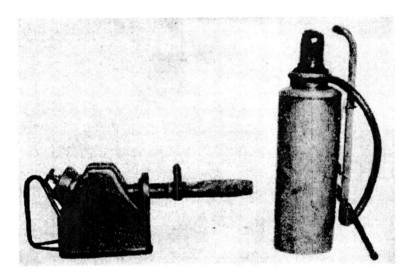

Fig. 46.—Alcohol torches suitable for hard-soldering where gas is not available.

If natural gas is to be used, be sure when you purchase your torch that it is of the type used with natural gas, for the torch made for illuminating (artificial) gas will not work satisfactorily on most natural gases. The Automaton, as well as a number of other torches, is also made specially for natural gases, but must be so ordered. The Hi-Heat torch will operate with either natural or artificial gas.

Alcohol Blowtorch. Alcohol or small gasoline blowtorches may be used in silver soldering when gas is not available. Although not so convenient as the gas torches, they are quite effective. The automatic alcohol torches are very good for soldering large pieces.

The alcohol torch shown at the left in Fig. 46 is an automatic one that produces a high heat. The one at the right is used by blowing through the mouthpiece.

An alcohol lamp and the conventional type of blowpipe may be used in making rings and small jewelry, but such equipment is not suitable for heavy soldering, as it is difficult to produce enough heat with it.

Outfits for Camps. For camp work and shops where gas is not available, a gasoline gas generator may be installed. It is simple to operate, weighs only a few pounds, and produces a gas vapor that is excellent for soldering.

A rotary blower, or foot bellows, must be used to force air through the generator, and a special blowpipe designed for use with gasoline vapor is needed.

Acetylene torches, using acetylene gas stored in tanks under pressure, are excellent. With these a pressure regulator should be used.

Silver Solder. Silver solder is sold by the ounce (troy) in sheets, wire, strips, and granulated form. Two general types are available, "easy" and "hard" flowing. The easy flowing is preferable for most jewelry work as it melts at a lower temperature.

Fig. 47.—Gasoline generator complete with foot bellows and torch. (***Buffalo Dental Mfg.*** Co., ***Buffalo, N. Y.***)

Handy & Harman, Wildberg Bros., and Dee & Co., all make a number of silver solders, which have different fusing points. Each solder has a specific use. For general jewelry work Handy & Harman recommend their grade sold as "Easy," which has a melting point of 1325°F.; Wildberg Bros. recommend their No. 3, which melts at 1375°F.; and Dee & Co. recommend Dee's No. 3, which melts at 1430°F. All three of these silver solders are silver white in color and a close match for sterling silver, as their silver content is high.

Although silver solder may be purchased in any desired thickness (Brown and Sharpe gauge) and then cut into pieces suitable for use, the author has found that 28 or 30 gauge in strips 1/16 inch wide is excellent for student use. Being thin and narrow, it is easily cut into small pieces. Some of the supply houses dealing in jewelry findings handle silver solder that has been cut into small pieces. If your solder oxidizes through exposure to air, before you use it, rub it with fine steel wool until it is bright.

A very small piece of silver solder should be used in soldering a joint. Do not use a piece large enough to form a high place on the work for this must later be removed by filing, a process that in many instances is difficult. It is easier, if necessary, to add more solder than it is to remove excess solder. On some joints, such as sections where ornamentation is soldered to a ring, it is almost impossible to remove excess solder. Most beginners are likely to use too much solder.

Preparing the Joint. Any joint to be silver soldered must first be scraped clean or filed and then made to fit snug, for this type of solder will not bridge a gap. If there is any likelihood of the pieces moving while being soldered, bind them with thin black binding wire, which is sold by supply houses for this purpose. Do not use copper, brass, silver, or any of the bright wires for binding purposes, as the solder will stick to them.

Whenever possible, solder without the use of binding wire, for sometimes the solder will flow alongside the wire and in some instances flow over it. When this happens, the ridge of solder must be removed by filing.

Fig. 48.—Applying solder to a ring shank-bezel joint.

Fluxes. A flux must be used on the joint to prevent oxides from forming when the work is heated. Powdered borax or stick borax dissolved to a saturated solution in hot water is a good flux. Borum Junk, obtainable from supply houses, ground in water to the consistency of a thin paste is excellent and is used extensively. Liquid fluxes for hard-soldering are obtainable and work satisfactorily. Battern's Self Pickling Flux, a liquid flux, sold by William Dixon, Inc., is good. It can be applied with a medicine dropper. Handy Flux, in paste form, sold by Handy & Harman in half-pound and larger jars, is excellent.

Apply the flux to the joint with a small artist's brush, being sure that the joint is thoroughly covered. Then dip the solder, which has been cut into small pieces, into the flux and apply to the joint with tweezers, brush, or toothpick.

The Craft of Jewellery Making

Fig. 49.—Student silver-soldering a ring, using mouth blowpipe and artificial gas.

Applying the Heat. With the work on a charcoal or asbestos block, apply heat gently until the water in the flux has evaporated. If heat is applied too strongly at first, the moisture in the flux will cause the flux to rise and thus to move the solder. Keep the flame in motion and off the joint as much as possible. When the joint shows a dull red, indicating that the temperature is around 1200°F., concentrate the heat upon the joint. The solder will flow when the parts to be joined reach the melting point of the solder.

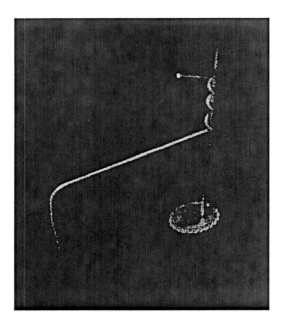

Fig. 50.—Method of holding scarf-pin stem in place on a charcoal block while it is being soldered.

Avoid using a small needlelike flame as this type of flame is very hot and is likely to melt the joint, especially on work on bezels and small wires. Very little air is required to produce the correct flame, especially if you are using artificial gas. If too much air is mixed with the gas, it will cool the work. Practice and experience will show what type of flame is best, as well as the correct amount of air required.

On large pieces such as bracelets it is well not only to place the work upon a charcoal or asbestos block, but to use charcoal or asbestos at the back of and on each side of the work so as to reflect the heat. Soldering should not be done in a draft as the air current will cool the work.

In soldering a small piece to a large piece, apply more heat to the large piece, for the solder will flow and join the two pieces only when both pieces are of the correct temperature.

Balling of Solder. If the solder rolls into a ball when it melts and refuses to flow readily, it is probably because the solder or joint was not covered with flux or because the work was underheated. Keep the flame upon the work as much as possible, instead of upon the solder; otherwise the solder will melt, ball, and will not flow readily when the work reaches the melting point of the solder.

Pinholes in the finished joint may be caused by dirt, oil, or improper fluxing, or by too much or not enough heat.

Soldering Ornaments. Small ornaments, shot, twisted wire, leaves, and other decorations, can be soldered onto larger pieces easily by the use of silver solder filings, known also as granulated silver solder. The filings are mixed with powdered Borum Junk, one part of solder being mixed with two parts of the powder. Using a pestle, crush the Borum Junk in a mortar and then sift it through a cloth. Mix a little at a time or, if a large amount is mixed, remove a small quantity and place it in a jar to be used while you are working.

Make sure, by pickling, that the ornament to be soldered in place is clean; then moisten the underside of the leaf or ornament and, holding it with tweezers, touch it to the filings-flux mixture. Put it in place and apply the heat, being careful to keep the flame off the ornament.

In soldering a small ornament, you will find that it will generally hold better if at the instant the solder melts the ornament is pressed against the surface with the tang end of an old file or a similar object.

Another method quite effective in soldering leaves to a ring or to other pieces of jewelry is first to put flux over the place on which the leaf is to be applied and then to glaze (melt) the flux. Next dip a small piece of the solder into a thick consistency of Borum Junk and water and place it on the underside of the leaf, usually in a hollow. Then apply the leaf to the object in the usual manner. When the work gets hot enough, the solder underneath the leaf will melt and flow onto both the leaf and the work.

Some craftsmen wet the tip of the tang end of a small discarded file, stick this into a mixture of silver solder filings and Borum Junk, and then touch the joint or piece to be soldered when it becomes red-hot. Borax must be applied to the joint before the heat is applied. This method is sometimes used in chain making, especially for chains made of gold wire.

Large joints can sometimes be silver soldered by applying flux to the joint and heating to a point that will melt the solder. A thin strip or wire of solder is then dipped in the flux and touched to the joint. This method is used extensively in work on long seams and joints.

Protecting Soldered Joints. When more than one joint is to be soldered, coat previously made joints with borax. If there is danger of the joints opening, a solder of lower melting point may be used, or the soldered joints may be coated with a paste made of yellow ocher and water, whiting and water, or jewelers' rouge and water. After the soldering is completed, the ocher may be removed by soaking the work in water and scrubbing with an old toothbrush.

All the jewelry shown in the illustrations in this book was made with one grade of solder, and very little difficulty was encountered. Yellow ocher was used on some pieces by beginners. Advanced students, as a rule, depend upon borax flux to hold the pieces together and to protect previously soldered joints, seldom using either binding wire or yellow ocher if their use can be avoided. Cautious application of heat in evaporating the water from the flux, so as not to disturb the solder or the joint, is essential.

Pickling. Glazed borax, as well as oxides left on the article after the joint is soldered, may be removed in a pickling bath prepared by adding 1 part sulphuric acid to about 15 parts water. *Add the acid to the water.*

The solution is more effective when used hot and should be heated in a copper pickling pan. When not in use, the liquid should be kept in a glass or earthen jar. Remove all iron binding wire and place the soldered article in the heated pickle bath. Retrieve the work with copper or silver wire or copper tweezers. Never use an iron wire or object to retrieve the work as the reaction of the iron in the solution will discolor silver.

Fig. 51.—Copper pickling pan.

If it is inconvenient to heat the acid solution, the article to be cleaned may be heated and, while hot, dropped into the cold liquid.

Making Silver Solder. Small amounts of silver solder may be made by fusing small pieces of silver and brass on a charcoal or asbestos block, the proportions being about 4 parts silver to 1 part brass. If a melting furnace is available, a larger amount may be prepared. Place the silver in a crucible, put in a small amount of borax, and melt the silver. Then add the brass filings and stir. Pour the alloy from the crucible and when cold make into granulated form by filing with a coarse file or hammer or roll into the desired thickness.

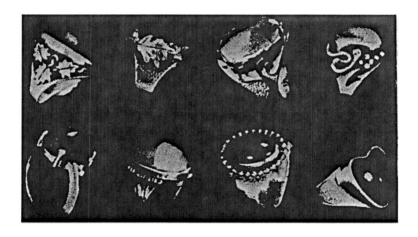

Fig. 52.—Sheet-silver rings using leaves, wire, and shot for decoration.

Silver solder, however, is not expensive and is best bought

ready-made.

Gold Solder. Gold solder is made of fine (pure) gold alloyed in varying degrees to melt at different temperatures for the different karat golds. Gold solder is used in the same way as silver solder, with borax as a flux. Owing to the fact that different karat golds have different melting points, a special solder for each karat gold must be used. Gold solder must also match in color. For example: 10-karat white gold is soldered with 10-karat white-gold solder, and 14-karat yellow gold is soldered with 14-karat yellow-gold solder. Gold solder is sold by the pennyweight (dwt.).

CHAIN MAKING

Before attempting to silver solder a ring or other piece of jewelry, one should practice first upon copper, which is inexpensive and works much the same as silver. A useful beginning exercise in hard-soldering is the making of a chain.

Secure a short length of 18- or 20-gauge round copper wire and a large nail. Wrap two layers of paper around the nail. Then hold the nail and the end of the wire in a pair of pliers and tightly wrap the wire around the nail. Hold the nail in a flame to burn out the paper. Remove the coil of wire from the nail and saw or cut into links. File the links until the ends fit snugly. Solder a number of links into single units. Then connect two such soldered units with a third unit, making units of three. When two units of three links each have been made, join these with another link, thus making seven units. Repeat until the desired length is obtained. Links of various shape can be made on mandrels of the desired shape cut from thick copper or sheet metal, or the links may be twisted by using pliers.

RING MAKING

Finger-ring making, the most popular form of jewelry work for students, may be divided into eight operations: (1) designing, (2) bezel making, (3) shank making, (4) assembling shank and bezel, (5) ornamenting, (6) polishing, (7) oxidizing, and (8) setting the stone.

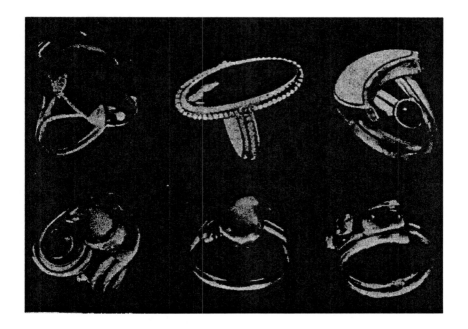

Fig. 53.—Wire shank rings.

Rings may be made from wire and sheet silver or from sheet silver only.

Designing. The designing of a ring must necessarily be based upon the ornament, or stone, that is to be used on the ring. Different kinds and shapes of stones require different mountings. Faceted and cabochon stones generally require different methods of ornamentation as well as of mounting. In designing a ring, one is limited only by his ability to execute his designs and by the material at hand.

Before starting work on the actual making of a ring, it is well to prepare a number of sketches of different types of mountings with decoration or ornamentation for each and decide definitely which is to be used. Often it is advisable to make these sketches larger than the actual size of the ring.

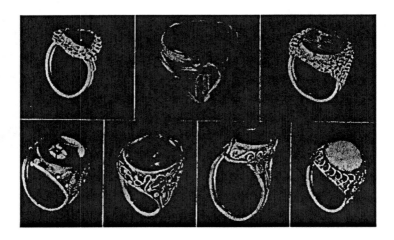

Fig. 54.—Rings ornamented with twisted wires and small balls of silver, termed "shot."

Bezel Making. Although either sterling or fine silver may be used in making the bezel, it has been the experience of the author that fine silver is preferable for student work, as it is much softer and will bend to shape more readily. Regular bezel made of sterling may be purchased, but it is harder to work than that made of fine silver.

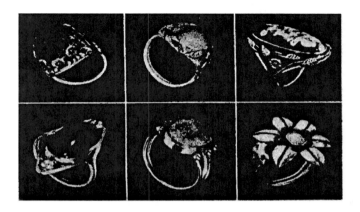

Fig. 55.—Rings using leaves and shot for decoration.

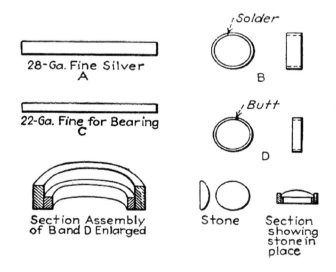

Fig. 56.—How the bezel is made.

The bezel consists of two parts, an outer rim that is burnished over at the top to hold the stone in place and an inner part, or bearing, that supports the stone and keeps it from going on through the outer rim. The outer rim is generally made of 26- or 28-gauge Brown and Sharpe fine silver. It is made to the exact shape of the stone, and the ends are silver soldered together. Fine silver may be purchased in widths suitable for bezel making, usually either 1/8 or 3/16 inch.

Fig. 57.—Rings.

If the outer rim of the bezel, when made, is too small, it may be stretched by slipping it over a round mandrel and hammering lightly. It may then be shaped to fit the stone. As the metal is quite soft and stretches easily, avoid heavy hammering.

The bearing may be made of small round sterling wire or of sterling or fine sheet silver cut into strips about 1/16 inch wide. Fine silver 1/16 inch wide may be purchased and then cut into the desired lengths. Although any of several gauges of silver may be used for the bearing, the 22-gauge Brown and Sharpe works satisfactorily. The bearing is made so that it fits exactly the inside of the outer rim. Do not solder together the ends of the bearing before inserting the bearing in the outer rim.

After the two parts are assembled, place them upon the soldering block and solder together.

Although fine silver has a melting point approximately 400°F. higher than that of the solder, it will quickly melt if too much heat is applied or if a pointed flame is used.

Shank Making. Shanks for rings may be made either from round wire or from sheet silver. Sterling silver is used for shanks, as the fine silver is too soft. If the shank is to be made from round wire, use 15- or 16-gauge Brown and Sharpe. Wire shanks may be made in either of two styles, as shown in Fig. 58. Secure two pieces of wire, about 3 inches in length, and anneal by heating to a dull red. Pickle and then bend the wires to shape. If desired, bind with binding wire to keep them from moving while they are being soldered. Use several small pieces of solder instead of one large piece.

After the wires have been soldered together, bend them to shape and solder to the bezel. The only difference in the two styles is that in type 1 the shank is made the size desired after it is soldered to the bezel, and in type 2 it must be sized before it is soldered to the bezel.

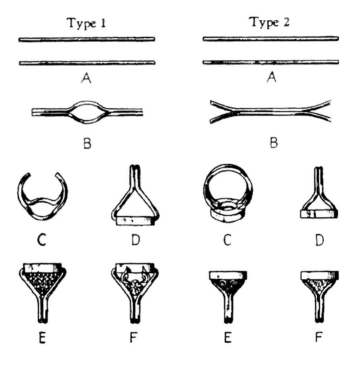

Fig. 58.—Steps in making rings from wire. *A*, 16-gauge round sterling wires. *B*, wires bent to fit bezel and soldered together. *C* and *D*, wires shaped and soldered to bezel. *E* and *F*, decoration applied to ring assembly.

Square or triangular wire may be used in making shanks. Shanks may also be made of sheet silver, 18- or 20-gauge Brown and Sharpe usually being used. In this case the bezel may be soldered upon a thin sheet of sterling and the shank soldered in place after being cut to shape and sized.

Assembling Shank and Bezel. After the shank and bezel have been made, it is necessary to solder them together. Place the bezel, inverted, upon the soldering block and put the shank in place.

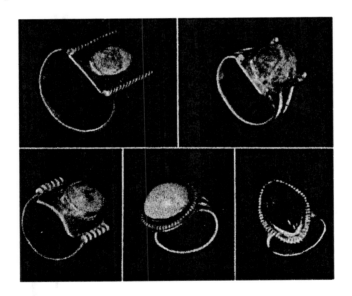

Fig. 59.—Rings made with sheet-silver shanks using twisted wire, shot, and beaded wire for ornamentation.

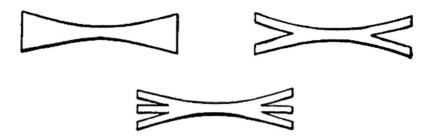

Fig. 60.—Shanks made from sheet silver.

If using a wire shank, type 1, the wires may be sprung apart just enough to insert the bezel, which usually will stay in place while being soldered. Both sides are soldered at the same time. If a wire shank, type 2, is used, it must be made of correct size and the end of each wire filed so that it fits against the bottom of the bezel. All four joints may be soldered at the same time. No binding is necessary if the flame is cautiously applied so that the borax will not "rise" to disturb the assembly.

Shanks made from sheet silver are shaped, sized, and placed upon the bezel or upon a thin sheet of sterling to which the bezel has been soldered, and both sides are soldered at the same time.

The Craft of Jewellery Making

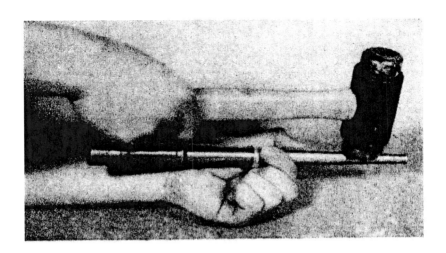

Fig. 61.—To remove a cabochon from a mounting, loosen the bezel around the stone with a penknife, place ring on mandrel, hold as shown, and strike end of mandrel a sharp blow.

Ornamenting. The ornamentation that is to be applied to a ring depends to a large extent upon the kind and shape of stone, as well as upon the kind of mounting, that is being used. One's ingenuity is, however, a large factor, as many variations and adaptations are possible.

The ornamentation usually consists of the following or combinations of the following: bent and twisted wires, shot, leaves, beaded wire, special designs cut from sheet silver, and saw piercing.

BENT AND TWISTED WIRES. Small wires, ranging in size from 20 to 30 gauge, may be twisted together and used for ornamentation. A good way to twist them is to tie the ends of two or more wires together, fasten on a hook, which is held in the chuck of the drill press, hold the free ends, and start the drill press. One can, however, twist them by hooking them onto any revolving arbor or by using a small hand drill.

The twist should be annealed before being used and then pickled. In annealing the twist, roll it into a coil about 2 inches in diameter and use a large flame of the torch with very little, if any, air.

Scrolls and various shapes may be made of the twists. The use of this type of decoration is shown in Fig. 54.

Instead of the wire being formed into scrolls, circles are sometimes made. If this type of decoration is to be used, the bezel is usually mounted upon a piece of 22-gauge sheet silver. The circles are made by wrapping the twisted wire around a small drill, usually a No. 55 or a No. 60. The coil is cut into links, which are soldered on the sheet silver. The drill is then used to bore out the metal. Small balls of silver, called shot, may be soldered onto the circles if desired.

Wire bent into many shapes or designs may be used to great advantage for decorating rings. Shot may also be used on this kind of decoration. If desired, the shot may be filed partly away to produce a flat effect. In using wire or any ornamentation that is applied to the bezel, do not get it too near the top of the bezel, or difficulty will be encountered in setting the stone.

SHOT. Small balls of silver, or shot, may be made by melting scraps of silver upon a soldering block. If, however, several balls of the same size are desired, equal lengths of wire may be cut and a ball made from each length.

To cut wires of equal length, wrap a piece of wire around a nail or other suitable mandrel and then cut the coil into single units. Always use borax on the silver when melting to form the balls. The shot should be pickled before it is soldered onto the ring.

Shot may be soldered in clusters, used at regular intervals, or used to fill in spaces that otherwise would appear open.

LEAVES. You can shape leaves from sheet silver by first sawing to the desired shape, cutting in the veins with an engraving tool or other sharp instrument, and then doming or dapping. The dapping is usually done on the end grain of wood or upon a lead block, with dapping punches, which may be purchased or made from tool steel. A large nail, the end of which has been ground to the desired shape and then polished, is excellent for emergency use when dapping punches are not available.

Commercial stamped leaves of various shapes and sizes may be secured and used in ornamenting.

If leaves instead of the bezel are to be used to hold the stone in place, make the bezel in the usual manner but make it very low, so that when the leaves are soldered in place they will project above the bezel. Solder each leaf on separately. After the ring has been cleaned, polished, and oxidized, the leaves are burnished over to hold the stone in place.

Fig. 62.—Leaf designs.

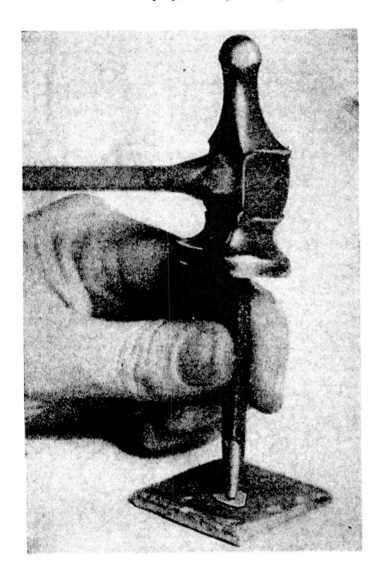

Fig. 63.—Making an ornament on the lead block.

BEADED WIRE. Beaded wire of various gauges may be secured from supply houses in any length desired, or it may be purchased by the ounce from a silver dealer. Beaded wire is available in round, half-round, and pearl bead. Often it is necessary to anneal beaded wire before bending it into shape. TRIANGULAR WIRE. Low-dome triangular wire is useful in jewelry making in a number of ways. The smaller sizes can be used in making ring shanks, as shown in the lower left ring in Fig. 52 and in the top center ring of Fig. 53, where two pieces of the wire were used and the V formed between the two was filled in with a beaded wire. Bracelets may also be made of the various sizes of triangular wire, as shown in Fig. 81, where each of the bracelets shown at the left was made from two pieces of the smaller sized wire with bead or twisted wire for decoration.

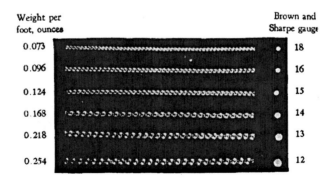

Fig. 64.—Round-bead wire. (*Courtesy of Wildberg Bros. Smelting and Refining Co., Los Angeles.*)

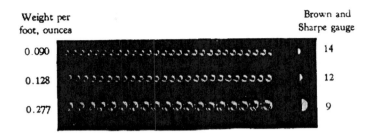

Fig. 65.—Half round-bead wire. (*Courtesy of Wildberg Bros. Smelting and Refining Co., Los Angeles.*)

SHEET-SILVER ORNAMENTS. Initials, monograms, Indian symbols, and many other designs may be sawed from sheet silver and soldered in place as ring decorations. The ring is usually made the correct size and the ends of the blank are soldered together before the ornaments are added.

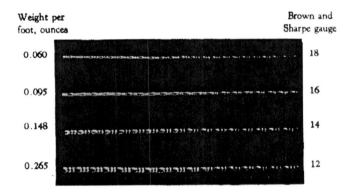

Fig. 66.—Pearl bead wire. (*Courtesy of Wildberg Bros. Smelting and Refining Co., Los Angeles.*)

The Craft of Jewellery Making

If the ornament needs to be curved to fit the ring, this may be done by making an impression in a wood or lead block with a ring mandrel and a wood mallet. The ornament is then put in this impression and the mandrel placed on top of it and again hit with the mallet. Be sure that the ornament is face down in the impression; otherwise the curve will be wrong.

SAW PIERCING. Saw piercing is often used on rings made from sheet silver and must be done before the blank is bent to shape. The piercing is done with a fine jewelers' saw, as explained elsewhere.

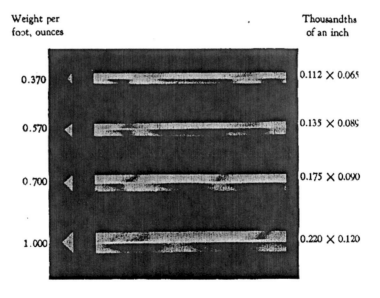

Fig. 67.—Low-dome triangular wire. (*Courtesy of Wildberg Bros. Smelting and Refining Co., Los Angeles.*)

POLISHING

After all silver soldering is completed, the ring must be cleaned in the pickling bath. Rough spots must be filed smooth, with jewelers' files, and the ring must be polished. The ring is polished with jewelers' rouge, used as an abrasive on a felt or muslin wheel, which is mounted upon a motor or arbor. Do not polish the ring upon a wheel that is used for polishing brass or copper. After polishing on the buffer, wash the ring with soap and hot water to remove the rouge.

Small scratches and file marks that are in places from which it is impossible to get them out with the buffing wheel may be removed by using water-of-Ayr (Scotch) stone. Keep the stone wet while rubbing. A slate pencil will serve almost as well, and fine steel wool or crocus cloth will remove many scratches. Use it before buffing with rouge.

OXIDIZING

A silver ring when polished will appear very bright. It will gradually turn darker if left exposed to air, especially if any sulphur is present in the air. It may be readily darkened by dipping into a solution made by dissolving a lump of liver of sulphur in a small jar of water. Reaction is better if the solution is hot. Because of the objectional odor of the liver of sulphur, a commercial oxidizing solution may be preferred.

This is usually applied with a small brush to the ring, while the ring is warm.

The oxidizing solution will discolor the entire ring. To remove part of the oxidization, a fine pumice powder, or a kitchen cleanser, mixed with water may be used. Rub it on with the fingers until the desired hue is obtained.

SETTING OF STONE

After the ring has been polished and oxidized, the stone is ready to be set. First taper the edge of the bezel with a file and then place the stone in the bezel.

Hold the ring in a ring clamp, and with a burnisher gradually turn the top edge of the bezel over onto the stone. This is best done by holding the burnisher in a position parallel to the base of the stone and going around the bezel, applying pressure, until the thin metal is turned in against the stone. The bezel may then be smoothed, if necessary, with a jewelers' file. Polish with a hand buff, which is a piece of felt glued to a thin wood strip, using rouge as an abrasive.

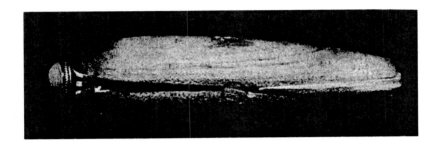

Fig. 68.—Ring clamp used for holding rings. Clamps like this are easily made.

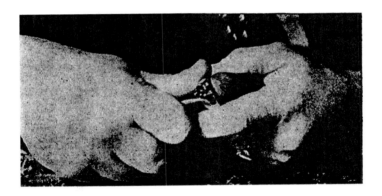

Fig. 69.—Setting a stone with a burnisher. The ring is held in a ring clamp.

The burnisher, or file, will not damage the majority of stones that are used, as the stones are harder than the file. Turquoise, variscite, malachite, opal, and a few others, however, being softer, can be damaged.

The Craft of Jewellery Making

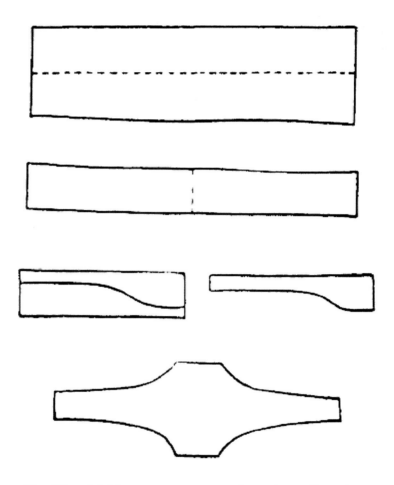

Fig. 70.—Making a paper pattern for a sheet-silver ring.

SHEET-SILVER RINGS

To make a ring from sheet silver, it is necessary first to make a paper pattern. If a stone is to be used, the width of the paper should be somewhat wider than the length of the stone. The length of the piece of paper is determined by the size of the ring. It is well to make a number of patterns and then to select the best. After the pattern for the ring blank has been selected, the design of the ring should be sketched on the pattern, especially if the decoration is to be saw pierced. One method of developing the paper pattern is shown in Fig. 70. Fold the paper lengthwise, then crosswise. Draw the outline on the folded paper, cut it out with scissors, and unfold.

Make the paper pattern the correct length for the ring size desired by measuring on a ring stick or upon the full-size scale, as shown in Fig. 39. Add the thickness of the metal to the correct length to take care of filing and bending.

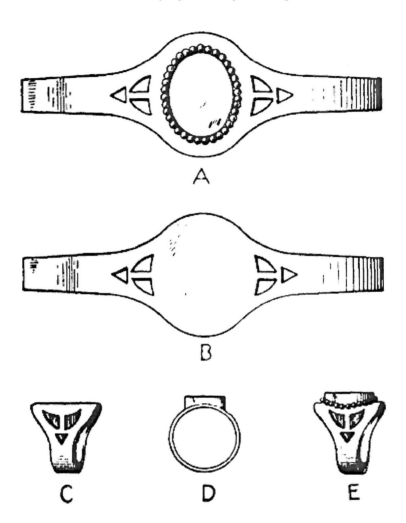

Fig. 71.—Steps in making a sheet-silver ring. *A*, paper pattern with design. *B*, blank sawed from sheet silver. *C*, blank sized and ends soldered together. *D*, bezel soldered in place. *E*, ornamentation soldered in place.

The Craft of Jewellery Making

Glue the paper pattern to the sheet silver, 16, 18, or 20 gauge Brown and Sharpe, and, using a jewelers' saw, saw out the silver blank. If designs are to be sawed into the blank, drill small holes, insert the blade, and saw out the design. Fine saw blades (No. 3/0 or 4/0) are preferable, for they leave a smoother cut. Saw over a notched block or a piece of wood fastened to a table top (Fig. 10). The blade cuts on the downward stroke. The method of applying tension to the blade as it is being inserted in the saw frame is shown in Fig. 9. Paraffin rubbed on the saw teeth aids in the sawing.

Ordinary glue will not adhere paper to metal satisfactorily. M.C. Glue, sold by Metal Crafts Supply Co., is excellent, and paper cement made from rubber is good.

Fig. 72.—Rings made from sheet silver using simple designs.

After the blank is sawed out and pierced, if designs are to be sawed into the blank, bend it to shape around a ring mandrel, using a wood or rawhide mallet. Then size the blank and solder the ends together.

Next make the bezel. Wider strips of silver than those used in making bezels for wire rings must be used, as the bezel must be cut away on the underside to fit the curved surface of the ring. Bezel material 3/16 inch and in some case 1/4 inch wide is used. The bearing or support is about 1/16 inch less in width.

To file the bezel to fit the ring, wrap one thickness of abrasive cloth around a ring mandrel and cut out the metal on the underside of the bezel by drawing the bezel back and forth over the cloth. At some place along the mandrel is the exact curvature of the ring, and when this is found and the cutting done there, the bezel can be cut away to fit the ring exactly. In order to avoid a tapering cut, change the ends of the bezel frequently during filing. Avoid mashing the bezel out of shape while filing. If desired, the stone may be kept in the bezel during the filing operation.

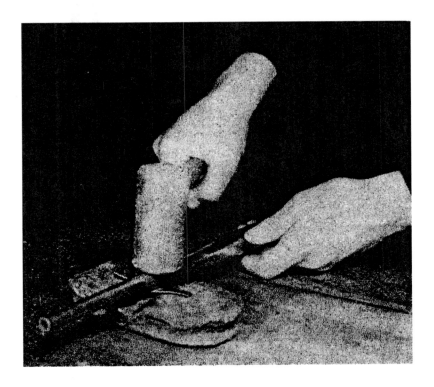

Fig. 73.—Use a mallet, ring mandrel, and lead or wood block to shape a sheet-silver ring blank.

The Craft of Jewellery Making

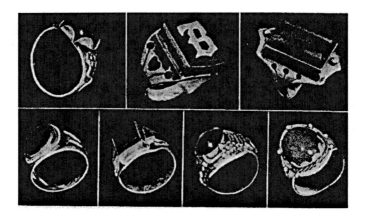

Fig. 74.—Sheet-silver rings with various kinds of ornamentation.

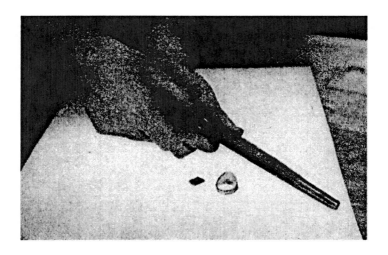

Fig. 75.—Filing a bezel to fit a sheet-silver ring on abrasive cloth wrapped around a ring mandrel.

To solder the bezel to the ring, coat the joint in the ring with borax flux and place the ring so that the bezel will be on top. Hold the ring in position with small pieces of charcoal or asbestos blocks. Place the bezel in position and coat the inside of the bezel as well as the joint of the bezel with a borax flux. Place several pieces of solder on the ring, inside the bezel, and apply heat gently until all water is evaporated. Heat should then be applied upon the side of the ring, below the bezel, from a rather large flame, until the ring is red-hot, at which time the flame may be directed upon the top of the ring and bezel.

If ornaments are to be added, solder them in place and clean, polish, oxidize, and set the stone in the usual manner.

RINGS FROM WIRE

Rings that are inexpensive, costing only a few cents each to make, and that appeal to many students may be made from various sizes of round sterling wire.

The knot ring is very easy to make. Two 4-inch lengths of wire, usually 16 gauge or heavier, are used. Tie the knot in one piece of wire, but before drawing the knot tight insert the other piece of wire through the knot and tie the knot in this piece. Draw both knots tight and solder the wires together. Then size the ring and solder the ends together.

Snake rings, especially the double-headed kind, are much harder to make than knot rings. The difficulty with the double-headed snake ring is to get the wire the correct length before bending to shape so that both heads will be visible when the finished ring is worn.

Fig. 76.—Knot ring made from round sterling wire.

Fig. 77.—Snake rings made from round sterling wire.

BELT BUCKLE RING

Of the many rings already described and shown, the belt buckle ring is perhaps the most popular with high school students. Girls, instead of wearing it as a ring, often wear it as a neckerchief slide.

The belt buckle ring is a filing, bending, and fitting project and is excellent for beginners as it requires no soldering.

This ring is made by cutting the blank from 20-gauge sterling sheet and filing it to shape. If a piece of silver 4 1/4 inches wide is used and the blanks sawed out as illustrated, there will be no waste, and the cost of each blank will be very little. The width of the master blank used in marking out the blanks is 3/16 inch for the narrow part and 3/8 inch for the wide part, from which the buckle part is formed.

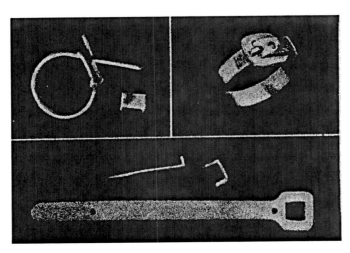

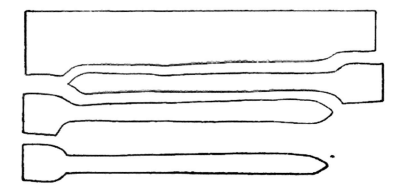

Fig. 78.—Belt buckle ring, showing method of assembling and how blanks are cut from a piece of sterling sheet.

The buckle part can be filed to any of a number of desired shapes. In filing the square hole in the buckle after it has been drilled, make sure that it is the exact width of the narrow part of the blank. The tongue and the U-shaped strap are made from small pieces of sheet silver. The method of assembling is shown in Fig. 78.

If desired, three holes may be bored in the band instead of one, as shown in the finished ring.

In assembling, make sure that the tongue goes between the buckle and the band, with the bent portion extending into the hole. The U-shaped strap is bent to shape with flat-nosed pliers. The ends of the strap are bent over to hold the assembly in place. All rough edges are filed off, and the entire ring is then buffed and polished. If desired, all joints may be hard-soldered.

Monel or nickel-silver may be used instead of sterling but is much harder to work.

BRACELETS

Bracelets may be made in a number of designs by the use of wire, sheet silver, and gem stones and may be decorated in various ways.

Four bracelets are shown in Fig. 79, made by different means. The one at the top has a stone mounted in the center of a piece of sterling sheet. Small wires twisted and formed into various shapes are used for ornamentation around the stone and on the six small pieces of sheet silver. Shot was soldered between the coils of wire.

The coin bracelet, made of souvenir coins, is simple in design and quite popular. Six coins are generally used, although five are sufficient. Small silver medals and commemorative pieces are sometimes substituted for the coins. Two of the pieces in the coin bracelet shown are not coins but are George Washington Bicentennial commemorative pieces. In selecting pieces for a bracelet of this type, choose those that have a high silver content. Coins from some countries have a low silver content and will, unfortunately, melt at about the same temperature as the silver solder, thus rendering them useless unless solder with a very low melting point is used. Coins that are unsuitable are generally yellow or brassy looking.

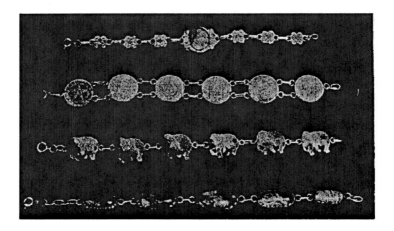

Fig. 79.—Bracelets.

The elephant bracelet is very similar in construction to the coin bracelet, except that the elephants are handmade from 20-gauge sheet silver. The blanks are first sawed out and the edges filed. Body lines and features are then cut with an engraving tool. The figures are made lifelike by being raised from the back. This is done by making a depression in a piece of lead or wood with a small ball peen hammer, placing the blank over the depression, holding the small hammer on the metal, and hitting lightly with another hammer.

The Craft of Jewellery Making

Fig. 80.—Bracelet made from three heart-shaped stones.

Many other designs may be worked into bracelets of this type. For instance, Indian symbols, such as thunderbirds, sun rays, arrowheads, and the Hopi horse, may be soldered upon sheet-silver blanks, cut into various shapes, and then linked together. If desired, a small turquoise may be set in each of the symbols.

Fig. 81.—Bracelets made from triangular wire.

Indian symbols may be suspended on chains, either handmade or commercial, to make charm bracelets. They are usually attached by soldering a small circle or loop to the top of the ornament and then connecting it with a link to the bracelet chain. Hearts, diamonds, spades, clubs, initials, monograms, crosses, and many other symbols may be attached to a chain to make a charm bracelet of pleasing design.

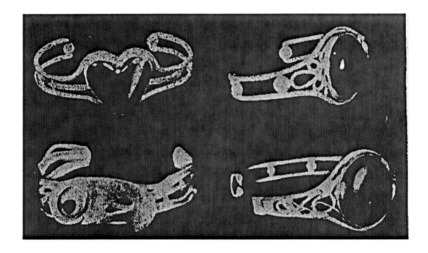

Fig. 82.—Bracelets.

The bracelet in Fig. 79 with five stones is only one of many designs that may be worked out where gem stones are featured. Three stones make a pleasing design if the central stone is larger than the other two, which are matched. With three stones some ornament is generally used in the bracelet between the large stone and each of the smaller stones. This ornament may be made of sheet silver or may be made of four small circles of 20-gauge wire soldered together to form a "square," with a large shot soldered in the central opening.

The spring ring used to fasten the bracelet around the arm may be purchased from supply houses or jewelry stores. The ring is usually attached by opening the small loop with a pair of pliers, inserting the chain, and then closing the loop with the pliers. Do not solder, since the heat would ruin the spring in the catch. Do not place the spring ring in the pickling solution, as the acid may cause the spring to rust.

Fig. 83.—Bracelets.

Four bracelets of the clamp-on style are shown in Fig. 83. Number 18-gauge sheet silver is ordinarily used for this type of bracelet, although the lighter weight 20 gauge is satisfactory. A piece of material of the desired width, about 5 1/2 inches long, is generally needed. Many different designs may be engraved, etched, stamped, or soldered onto the band for ornamentation, or stones in combination with any of the above methods may be used.

Here again are many possibilities for the use of the Indian symbols. One of the bracelets in Fig. 83 uses three gem stones, amazonites, in connection with Indian symbols, thunderbirds, cut from sheet silver. Steel dies of Indian symbols were used for further decoration on this bracelet. The design of the thunderbird track was stamped alongside each stone set on the thunderbirds, and the symbol of the sun rays was stamped around the central stone. Symbols of the arrow and the sun rays were cut from sheet silver and used in ornamenting the bracelet where a turquoise was set in the center.

FIG. 84.—Bracelets.

Other bracelets of this type using Indian symbols include narrow bands with the various symbols stamped into the metal. Raised hogan designs are especially useful for ornamenting, as are those of the arrow, the sun rays, and the thunderbird track.

Various adaptations of the thunderbird may be sawed from sheet silver and soldered onto the bracelet as the central object of ornamentation.

BANGLE BRACELETS

Bangle bracelets, a popular project, are made from 12-gauge, or heavier, round sterling wire. For an arm of average size a piece 7 3/4 to 8 1/4 inches in length is sufficient for one band.

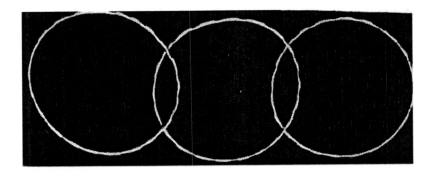

Fig. 85.—Bangle bracelets made from 12-gauge round sterling wire.

The ends of each piece are filed square, then soldered together. The bracelet is then placed over a round piece of iron and hammered with a wood or rawhide mallet to make it round. All hammering is done at one spot on the iron, and the bracelet is moved after each hammer blow. The iron does not have to be the size of the bracelet.

For decoration the band may be hammered with a small ball peen hammer, or flat indentations may be made at regular intervals.

Bangle bracelets may be made by twisting two or more wires. For a single twist use 20 inches of 12-gauge, or smaller, wire. Bend the wire at the middle, place the two ends in a vise, and, using a nail in the loop, twist the strands. If desired, run the twist through a rolling machine to flatten it. Use 8 1/4 inches of the twisted wire to make an average-sized bracelet.

The bangle bracelet shown at the upper left (Fig. 84), was made by twisting two wires, rolling the twist, and then soldering a wire on each side of the twist. If desired, the side wires may be pulled through a square drawplate. Flatten the sides of the twist with a flat file before soldering the wires together. Solder the wires together before bending. Use binding wire, and place a tiny piece of solder at each place at which the twist touches the outside wires.

BELT BUCKLE BRACELET

A bracelet that is popular with many high school girls is the belt buckle bracelet, which may be made of monel metal or of sterling. It is very similar to the belt buckle ring, except that it does not form a continuous circle.

The bracelet may be made in any desired width, although the width of the band is usually 1/2 inch and that of the buckle part is 1 inch. If the bracelet is hammered with a ball peen hammer, a pleasing effect is obtained.

Fig. 86.—Decorating a bangle bracelet.

The Craft of Jewellery Making

Fig. 87.—Belt buckle bracelet.

NECKLACES AND PENDANTS

The number of designs in which necklaces and pendants may be made is limited only by the material at hand and one's own ingenuity.

Fig. 88.—Pendants.

The necklace in Fig. 34 is made of garnets, both faceted and cabochon, ornamented with leaves and twisted wire. The heavy wire forming the outline of the heart was shaped and smaller wires carefully worked into the design so that they held the bezels of the four heart-shaped cabochons and the five faceted stones. The leaves were then soldered to the wires.

To form the chain of faceted stones linked together, first the required number of bezels were made, with rings soldered on at each end. The bezels were then connected with links. In work of this type no stone is set until all soldering is completed and the mountings cleaned, polished, and oxidized.

Fig. 89.—Drilled pendants are always popular.

The Craft of Jewellery Making

Fig. 90.—Pendants.

Four different pendants are shown in Fig. 88. In No. 1 one oval and three round cabochons are used; No. 2 uses a single stone, a moss agate ground into a heart shape, with the same curvature on both sides. No bearing is needed in setting a stone of this type. A narrow band of silver is shaped to fit the stone, a ring soldered in place, the stone inserted, and the edges of the silver burnished over against the stone.

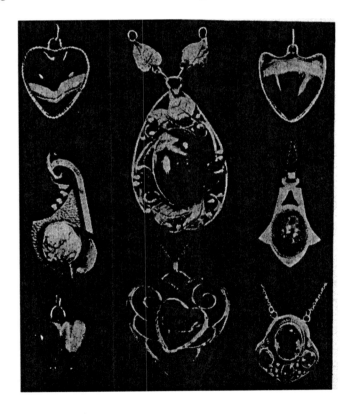

Fig. 91.—Pendants.

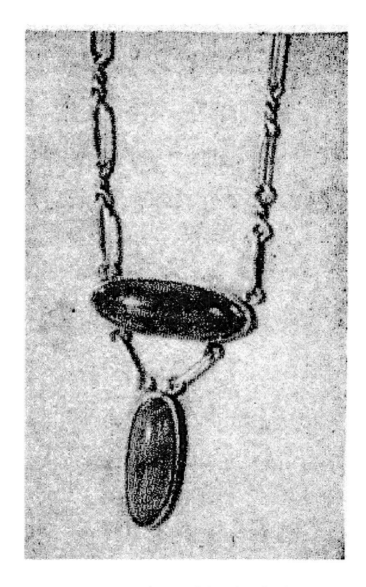

Fig. 92.—Pendant with handmade chain.

The Craft of Jewellery Making

Fig. 93.—Pendants using twisted wire for decoration.
(*Courtesy of N. Mardirosian, New York City.*)

Similar pendants may be made by using heart-shaped cabochons that are flat on one side (Fig. 91). Use the conventional-style bezel.

A large amethyst, surrounded by garnets and leaves, is used in No. 4, Fig. 88. After the bezel for the large stone had been made in the usual manner, a wire was soldered on the outside of the bezel. Another wire was shaped so that the small bezels might be soldered onto it and the wire on the outside of the bezel. The leaves and shot were then soldered in the spaces between the small bezels.

Twisted wires and shot were used in decorating the pendant

using four amethysts mounted upon sheet silver, as shown at lower right in Fig. 91.

Many simple designs may be sawed from sheet silver, such as the cross in No. 3, Fig. 88. A circle is soldered in place to hold the link through which the chain is run. A number of Indian symbols, sometimes set with small turquoise, may be used for pendants.

A chain 18 or 20 inches long is generally used for necklaces and pendants. This may be purchased with the catch already attached, from dealers, or the chain may be purchased by the foot and cut to the right length and the spring ring attached. Instead of a spring ring, a fastener made from a short piece of wire may be used. A ring is usually soldered to the middle of the wire, and the chain attached to this ring. A large loop is needed on the other end of the chain, through which the wire, or bar, is inserted. If desired, a ball may be soldered on each end of the bar.

SCARF PINS

The making of a scarf pin requires no processes other than those already explained. A scarf pin consists generally of two pieces, the mounting for the stone and the pin stem.

The method of holding a pin stem in place while soldering is shown in Fig. 50. Pin stems may be purchased with or without a twist in the stem.

The Craft of Jewellery Making

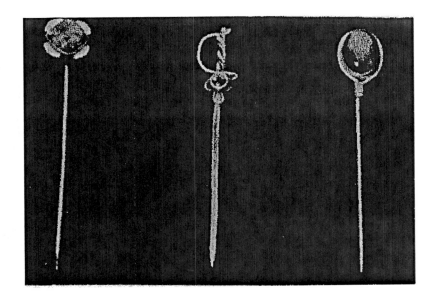

Fig. 94.—Scarf pins.

The effect obtained in the pin at No. 1, Fig. 94, is produced by mounting the bezel upon a piece of sheet silver that has been sawed to the desired shape.

The sword effect in No. 2 is obtained by soldering several pieces of wire and sheet silver together. The blade of the sword, which serves as the pin stem, was made of 20-gauge sheet silver. A small faceted stone is mounted on the hilt near the guard. The blade of the sword is made stiff by hammering.

In No. 3 a stone is mounted in a bezel in the usual manner. The pin stem is made of 17-gauge round sterling wire. One end of the wire goes around the bezel.

Pin stems for hard-soldering are usually straight. After the stem is soldered in place, it is bent to the desired shape with pliers. Pin stems, already bent, with patches riveted on, are obtainable for soft-soldering.

Fig. 95.—Cuff links.

CUFF LINKS

There are several methods of mounting gem stones for making cuff links. The method shown in Fig. 95 is quite satisfactory, however, for most stones.

The bezel is made and soldered onto a piece of sheet silver. A loop made of half-round wire, or round wire filed flat at points of contact, is made and soldered to each half of the cuff link. A link is then used to connect the halves.

EARDROPS

By the use of pierceless ear wires, many types of eardrops may be made with both faceted and cabochon stones. The eardrops shown in Fig. 96 were made of amethysts. The bezels for the large round stones were mounted upon sheet silver and then soldered to the ear wires, after which the decoration was applied.

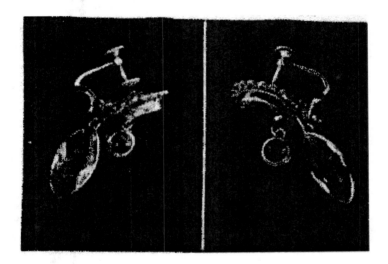

Fig. 96.—Eardrops.

The stones used in the eardrops in Fig. 97 were also mounted upon sheet silver. The decoration is shot of uniform size soldered to the sheet silver. The connecting link is made of wire.

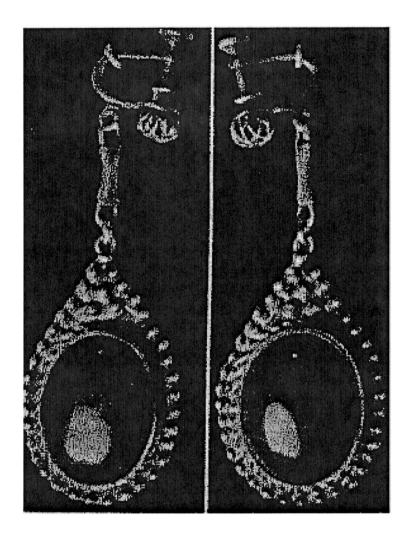

Fig. 97.—Eardrops. (*Courtesy of N. Mardirosian, New York City.*)

TIE CLIPS

With sheet silver and wire, many types and designs of tie clips may be made, with and without stones for ornaments.

Fig. 98.—Tie clip.

The wire clamp is made of 14-gauge round sterling wire. About 11 inches of wire is required. After all soldering and polishing have been completed, the wire, before it is folded over, is hammered stiff. Nickel silver tie clasps are available from dealers and can be adorned with handmade chains and ornamentation.

Spring tie clips may be secured and soft-soldered onto the underside of various ornaments.

BROOCHES AND CLIPS

Brooches and clips may be made of sheet silver and wire and ornamented by any of the methods described under ring making. Bezels for the gem stones are made in the same manner as those for rings.

The bezels for the three small turquoise used in the handmade flowers, Fig. 100, were made by twisting three strands of 30-gauge wire and then running the twist through a flat rolling mill. If no rolling mill is available, hammer lightly. No bearing or inner ring is required to support the stone. If the stone is not high enough, a small piece of sheet silver may be placed underneath it. The bezel, being made of twisted wire, has a pleasing effect when burnished over against the stone.

The brooch shown at lower center, Fig. 100, was made by mounting the bezels upon thin sheet silver and sawing out the portion inside the bezel. Twisted wire and shot were then used to fill in between the small bezels.

Catches, both plain and safety, as well as joints, are obtainable on patches, which can be tinned with soft solder. They are sweated into place. A much better, stronger joint may be made by obtaining catches and joints made for hard-soldering. The catch, or joint, is put in position, a piece of solder placed beside it, and the brooch heated until the solder flows, soldering the catch to the brooch.

Another method, preferred by many students, is to place a piece of solder covered with flux on the brooch at the spot where the catch or joint is to be attached and to heat the brooch until the solder melts. The base of the catch is then coated with flux and held upon the brooch, which is still hot, until the water has evaporated. The catch will usually stay in place when the brooch is again heated and the solder melted. Keep the flame off the catch, for if the catch is heated too much, the solder is likely to flow onto it, rendering it useless.

The Craft of Jewellery Making

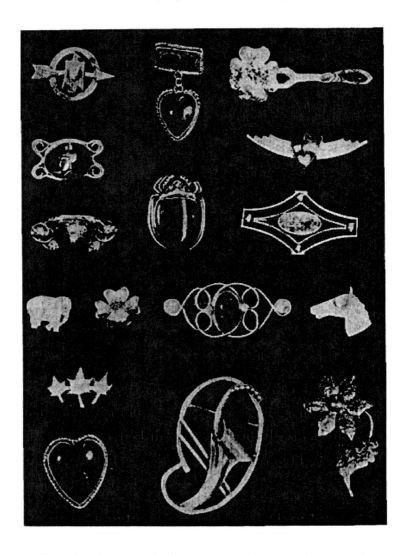

Fig. 99.—Pins made from sterling sheet and wire and inexpensive gem stones.

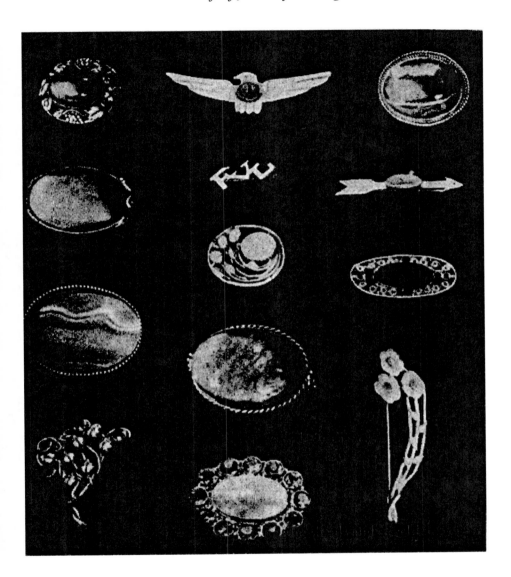

Fig. 100.—Brooches.

Both joint and catch are usually soldered above the center of the brooch, so that the brooch will hang better. The joint is soldered on the right-hand side, in relation to the wearer. The catch is soldered on with the opening toward the bottom of the brooch.

Fig. 101.—Brooches and clips. (*Courtesy of N. Mardirosian, New York City.*)

The pin stem is riveted to the joint by the use of a small nickel wire, called rivet wire, which is available in assorted sizes. Silver wire will, however, serve the purpose. Pin stems with the rivet in place may be obtained although they require a special joint made for this type of pin stem.

The design of the top center brooch, Fig. 101, is unusual in the handling of both leaves and ornament. The leaves are of twisted wire coiled into shape. The ornament, a fly, in the center of the carved carnelian stone is of sterling, with the body made of a tigereye gem stone. The fly is mounted upon a silver tube, which extends through a hole drilled in the carnelian, and is burnished over on the underside to hold the ornament in place.

Fig. 102.—Costume ornament.

Clips may be made with a solid back of sheet silver or with the conventional-style bezel, which is open on the back. The spring clips used on the back of dress clips must be taken apart before being soldered into place, in order to protect the spring.

Various ornaments may be made that are to be sewed to the clothing, like the ornament shown in Fig. 102. Initials and monograms may also be cut out of sheet silver and attached to ladies' purses. Small wires, which are to be pushed through the material to which the initial is to be attached and bent over on the underside, are generally soldered to the back of the initial.

BELT CHAINS AND WATCH FOBS

Belt chains may be constructed by making the belt loop out of sterling sheet and attaching the chain and the swivel. The belt loop or slide is made of one piece of silver, bent to shape and the end soldered. One end ordinarily projects, through which a hole is drilled to attach the chain.

Watch fobs are usually a sawing project, to which initials, monograms, gem stones, or some other desired ornaments are soldered.

KEY CHAIN

The key chain shown in Fig. 103 is made of three parts: the "horseshoe," by means of which the key is slipped on the chain, the ornament, and the connecting chain.

Fig. 103.—Key chain.

The horseshoe may be sawed from 18-gauge sheet silver. A hole is drilled in it, as shown, to attach the chain, which is usually about 2 inches in length. The ornament at the other end of the chain may be a gem stone, souvenir coin, or any appropriate design, such as a monogram, soldered to a silver blank.

The stone used in the illustration is a bloodstone intaglio. An intaglio is a stone with a face or design cut into the stone. Many intaglios in both bloodstone and sard are ground with a groove in the edge. If such a stone is used, the bezel is a piece of 20-gauge wire slightly longer than the circumference, or girdle, of the stone. It is run through the chain link, and the ends are soldered together. It is then placed around the stone and the stone set by twisting the wire loop.

INDIAN JEWELRY

When one thinks of Indian jewelry today, he almost invariably thinks of the Navajo Indians of the Southwest, who long ago became artisans in the use of silver, decorating most of their jewelry with turquoise.

Mexican pesos were used for many years as the source of supply for silver by the Indians, but today silver alloyed to .900 fine, known as coin silver, is generally used.

The nomadic Indian, herding sheep over the barren areas and moving from place to place, had to make use of simple tools to fashion his silver. These usually consisted of a hammer and a piece of iron upon which to hammer. For heat he used charcoal and a hand bellows. In soldering he used silver dust and alum moistened with saliva.

Genuine Navajo dies were fashioned by each silversmith

from pieces of iron and, being soft, could not be used to stamp the imprints into the silver. Instead the dies were pressed into the silver while the silver was red-hot.

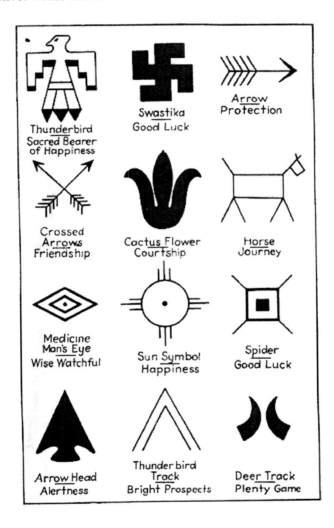

Fig. 104.—Indian symbols.

The Craft of Jewellery Making

Turquoise, which from the dawn of civilization has always been held in high esteem, has long occupied an important place in the mythology and folklore of the Indians of the Southwest. Marvelous virtues have been attributed to it: "One who sees turquoise early in the morning will pass a fortunate day"; "The eye is strengthened by looking at a turquoise"; "The turquoise helps its owner to victory over his enemies, protects him against injury, and makes him liked by all men."

Much of the present-day Indian jewelry that is bought at trading posts and gift shops is made by Indians under the supervision of the white man. It is stamped out on presses, with an Indian operating the press, and taken to an Indian silversmith, who solders on the cups or bezels, which are often factory made. The Indian is allowed very little use of his own ingenuity.

Indian-type jewelry is quite popular with high school students. Many of the Indian symbols can be used in decorating handmade jewelry. They can be cut from sheet silver and soldered to rings, bracelets, and other pieces of jewelry.

Steel dies of Navajo design, to be used in stamping Indian designs into silver, may be purchased and used in ornamenting handmade jewelry.

DISCARDED DENTAL INSTRUMENTS

Many of the instruments and tools used by dentists are very

useful in the fashioning of handmade jewelry by the home craftsman. Dentists generally discard scalers, burnishers, excavators, enamel cleavers, and similar instruments, as they break or wear down through use or frequent sharpenings. A discarded instrument can be readily ground on a grinding wheel to a shape or point that will prove very useful in jewelry making.

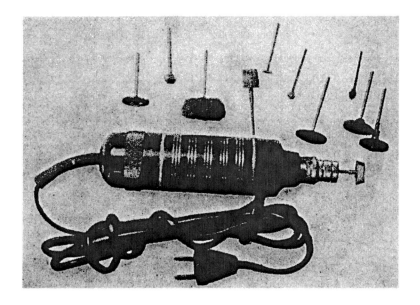

Fig. 105.—Hand motor tool and accessories.

Since dental instruments are generally made of the finest type of steel or one of the many chrome-steel alloys, even discarded instruments will give long and useful service in jewelry making. A broken scaler or chisel, for instance, can be sharpened to a point and will make a hand punch or scratch awl. Burnishing down of small bezels can also be accomplished by many of the dental instruments.

The hand electric motor tool will serve admirably for holding small grinding wheels, mandrels, sandpaper disks, abrasive wheels, and polishing buffs.

After the assembling and soldering of a ring or ornament, there are generally rough spots that should be ground away prior to polishing, and this can be easily accomplished with the above equipment. The abrasive disks will be found more serviceable than files in many instances. Polishing can also be carried out with the small felt buffs about the size of a five-cent piece, with cake rouge used as an abrasive. Small muslin buffs about 1 1/4 inches in diameter can be made by sewing a number of pieces of muslin together and mounting these upon a small mandrel that fits the chucks already referred to. The claw type of rings are especially difficult to finish and polish by ordinary means but are a simple problem with the dental equipment.

USE OF INVESTMENTS

In soldering a large or complicated piece of work made of gold or silver, it is often desirable to protect portions of the work where soldering has already been done. The joints may be protected by yellow ocher, as previously described, or with the investments used by dentists. These materials are termed bridge investments and consist essentially of finely shredded asbestos and plaster of Paris. Work protected with them will not come unsoldered through fusion during the soldering of the portions of the work left exposed.

Investments can be made easily at little expense by mixing 1 part of fine powered or shredded fibrous asbestos with 2 parts of ordinary plaster of Paris. This material is mixed with water and handled like plaster of Paris. It will set in about 5 minutes and then can immediately be heated by the blowpipe, without cracking and with little shrinkage. Plaster of Paris alone would tend to fracture under high heat. In applying the investment to the part of the work to be protected during soldering, mix the investment material with water to a fairly thick mass and apply, leaving exposed the parts to be soldered. After the investment has set for 5 minutes, the excess can be trimmed off with a knife prior to heating and soldering. Do not leave a huge bulk of investment material attached to the work, as this will only absorb an excessive amount of heat. Only a thin layer of investment is needed to prevent any part of the work from "burning up." The large and complicated gold bridges made by dentists are assembled by the use of investments.

RING CASTING IN SAND

A flask, some fine casting sand, glycerine, parting sand or powder, a crucible, a melting furnace, and models are the requirements for casting rings and other objects in silver.

Making the Flask. Although the flask shown in the accompanying illustrations was made from two 1 1/2-inch lengths of 3 1/2-inch-inside-diameter brass tubing, other sizes of tubing or pipe or wood frames could be used equally well. First, cut off the tubing, or pipe, to the required lengths and file smooth. Then, using a piece of copper, steel, or brass rod, shape the two eyes, or ears, and soft-solder them to one of the pieces of tubing. Next bore a 3/16-inch hole in each ear, being sure that the hole just touches the piece of tubing. The half of the flask with the ears is known as the drag; the half with the pins is known as the cope.

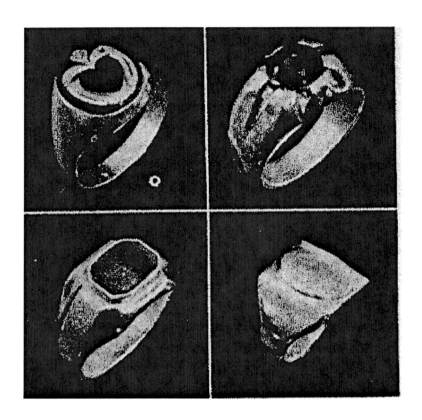

Fig. 106.—Rings cast from silver.

Fig. 107.—Drag ready to receive sand.

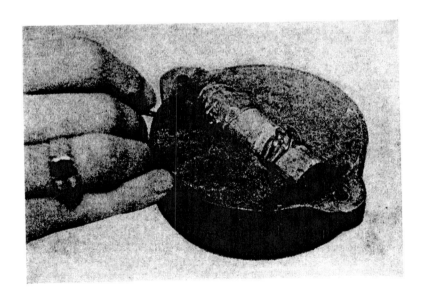

Fig. 108.—Repairing a mold.

Put the two pieces of tubing in a vise and insert short pieces of 3/16-inch rods through the holes in the ears and, after aligning, soft-solder them in place. If the rods fit too tightly in the holes, reduce their size with a file and smooth with abrasive cloth. Flasks for ring casting may, however, be purchased ready for use.

Preparing the Sand. Fine casting sand is needed for silver casting. In preparing the sand for use, sift through a sieve made of screen wire. Then work in just enough glycerine to hold the sand together, mixing both thoroughly. After each use the sand should be sifted, and after repeated use, if the sand becomes dry, work in more glycerine. Satisfactory commercial sand for silver casting may be purchased, ready for use, in 5-pound cans.

The Craft of Jewellery Making

Fig. 109.—Mandrel and sprue pins removed.

Mandrels and Core Tube. Secure a 3-inch length of copper or brass tubing that has an inside diameter equal to the size of the ring to be cast. Then, on a lathe, turn a wood dowel about 12 inches long, which will just slip inside the tubing.

Cut off two pieces of the dowel the exact length of the piece of tubing. Take one of these pieces and split it in half lengthwise, and mount it upon a 5- by 5-inch board, as shown in Fig. 107. Use the other short piece of dowel for the casting mandrel to hold the models, as shown in Figs. 108 and 110. Use the long piece of dowel to push the sand core out of the small tubing. Different-sized mandrels and core tubing must be used for each size of ring.

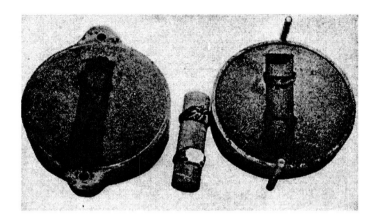

Fig. 110.—Mandrel and patterns removed.

Models. Before a ring can be cast, a model or pattern is necessary. It may be made of wood, soft solder, or plastic wood. If plastic wood is used, the models may be formed around the mandrel.

Maple or any other hard, close-grained wood is excellent for making ring models. Through a piece of the wood, bore a hole as nearly the size of the mandrel as possible and then file it out until it fits the mandrel snugly. Saw out the wood to approximately the shape and size of the finished model and finish with files and sandpaper. After the model is completed, give it and the mandrels a coat of lacquer or shellac and rub them down with steel wool until smooth.

In the making of models avoid undercutting, for the sand will not leave the pattern properly unless the pattern has a gentle slope from the bottom toward the top or, in the case of a ring, from the inside toward the outside.

Sprue Pins. Tapering plugs, called sprue pins, are needed to make holes in the sand through which the molten metal can reach the mold. These holes are known as sprue holes. After making the pins from either wood or metal, bore small holes in the half mandrel to hold them in an upright position, as shown in Fig. 107.

Preparing the Mold. In preparing for a cast, place the tapering plugs (sprue pins) in the half mandrel, dust both mandrel and plugs with the parting powder, and place the eye half of the flask, the drag, over the mandrel. Fill with sifted sand, packing lightly. Remove the pins, invert the flask, lift off the board, and you have what is shown in Fig. 109.

Place the model, or models if two rings are to be cast, on the round mandrel, dust with parting powder, and place the mandrel in the sand, pressing it down firmly so that the model will make an imprint in the sand. If any sand is disturbed, it may be repacked with a small spatula.

Dust the sand, mandrel, and patterns with parting powder. Place the pin half of the flask in position and pack it with sifted sand. After this has been done, carefully pull the two halves of the flask apart and remove the mandrels and patterns, as shown in Fig. 110. If any sand falls into the mold, it should be carefully removed.

Making the Core. Using the piece of tubing with the inside diameter the same as that of the model, pack it with casting sand and push out with a dowel (Fig. 111). If it is correctly packed and if care is used in removing, the sand will come out in a cylinder, or core. Place this carefully in the pin half of the flask and put the eye half of the flask in position.

Fig. 111.—Removing the sand core from tubing.

Melting the Silver. Place some silver in a graphite crucible and sprinkle a little powdered borax over the top of the silver. Place the crucible in the melting furnace, which is shown in Fig. 1. This type of furnace requires constant air pressure, which is usually supplied by a rotary air blower (Fig. 44).

After the silver has melted, remove the crucible with a pair of tongs and pour the molten metal into the holes left in the sand by the sprue pins. Allow the metal to cool for a few minutes and then pull the flask apart. If the cast is satisfactory, the ring, when cool, may be sawed from the tapering piece of silver formed in the sprue hole. It may then be finished by filing, sanding with abrasive cloth, and, finally, buffing. If for any reason the cast is unsatisfactory, the silver may be remelted.

Fig. 112.—Crucible and flask after pouring.

Metal Models. If a large number of casts are to be made from a model, it is advisable to prepare a metal model from the wood model, as the wood model is likely to break. The metal model is also superior in that it can be made to more exacting dimensions and will make a much cleaner cast that needs much less filing.

To make the metal model, instead of pouring silver into the mold, pour molten brass. Then file and buff the brass model until it is smooth. As the cast-ring model will be a little smaller than the wood model, it will not fit the mandrel unless it is stretched or sawed through the shank. The sawing is preferable, for then the model will fit several sizes of mandrels.

The gap left between the sawed ends will not interfere with the casting. When the impression is made in the sand, this irregularity may be removed with a small spatula or by inserting a finished cast ring in the mold and gently pressing it into place.

CASTING FLAT OBJECTS

Small flat objects may be cast very easily. Make a pattern and then place the pattern and the eye half of the flask on a board. Pack with sand, holding the sprue pin in place. Remove the pin, invert the flask, lift off the board, but leave the pattern in place. Dust with parting powder and place the pin end of the flask in place and fill with sand. Invert the flask, separate the halves, and remove the pattern. Put the flask together again, and it is ready for pouring.

FINISHING CAST RINGS

Rings that are cast in sand may be set with stones of various kinds, or initials or emblems may be soldered or engraved upon the various types.

Several different styles of cast rings are shown in Fig. 106. Three of them are designed either for initials, emblems, or stones; and one is designed for faceted stones. The bezel for the stone set on top of the ring was made of fine silver and soldered to the ring. After soldering a bezel to a cast ring, avoid sudden cooling of the ring, as it may warp. The carved agate is set flush with the top of the ring. This is accomplished by cutting out the metal and leaving a narrow rim, which is burnished over the stone and holds it in place.

The faceted stones are somewhat harder to set as the bearing must be cut on an angle. This may be done by filing, but a quicker, as well as an easier, way is to use a bearing burr. The method of cutting the bearing with a burr is shown in Fig. 113. The ring is held in a wood clamp. A hole is bored through the ring with a drill that is a little smaller than the stone. The burr is then inserted in the hand drill, and a seat or bearing is cut for the stone.

After the bearing is cut, the prongs are filed thin at the top, the stone set in place, and by using a burnisher the prongs are forced over the edge of the stone to hold it firmly. Bearing burrs may be purchased singly or in sets. A set of 30 different-sized burrs is shown in Fig. 114.

GOLD AND SILVER

Troy weight is used in weighing precious metals—gold, silver, and platinum.

TROY WEIGHT
24 grains = 1 pennyweight (dwt.)
20 pennyweight = 1 ounce (oz.)
12 ounces = 1 pound (lb.)

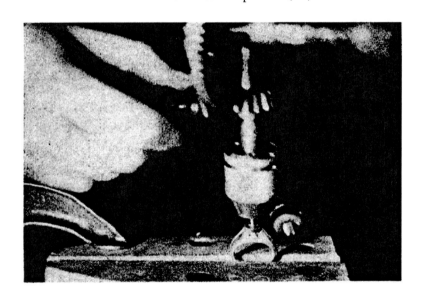

Fig. 113.—Cutting a bearing for a round faceted stone.

The Craft of Jewellery Making

Fig. 114.—Bearing burrs.

The troy pound contains 5,760 grains; a pound avoirdupois weight contains 7,000 grains. The troy ounce is about 10 per cent heavier than the avoirdupois ounce. Gold is sold by the pennyweight; silver is sold by the ounce.

Fine Gold. Fine (pure) gold is known as 24-karat gold. It is very soft and is rarely used for jewelry purposes. The melting point of 24-karat gold is 1945°F.

Colored Golds. Fine gold is alloyed with silver, nickel, and zinc in varying degrees to produce different colors and to make it harder. If an article is stamped 18 karat, the fine gold in the article is 18/24 the total weight. Likewise, if it is stamped 10 karat, it is 10/24 fine gold by weight. (Various manufacturing laws allow a slight deviation to compensate for the solder used.)

The component metals used in various colored golds are as follows:

Yellow golds for general purposes: gold, silver, copper, small amount of zinc.

Yellow golds for enameling purposes: gold, silver, copper.

Green gold for general use: gold, silver, small amount of copper, small amount of zinc.

Green gold for enameling purposes: gold, silver, small amount of copper.

White gold for general use: gold, nickel, small amount of zinc.

Red gold for general use: gold, copper, small amount of silver.

Melting Points of Golds. The approximate melting points of various karat golds are as follows:

	Degrees Fahrenheit
10-karat gold	1480
14-karat gold	1550
18-karat yellow gold	1640
18-karat white gold	2015

All white golds melt at a higher temperature than yellow, red, or green golds because of the nickel that is added. Different 10-, 14-, and 18-karat golds vary as to melting points because of the variance of alloys that are generally used to obtain certain colors.

Gold Filled and Rolled Gold Plate. Gold filled is made by joining a layer of gold alloy to a base metal and then rolling or drawing to the required thickness.

Rolled gold plate is the same as gold filled, although usually of a lower quality.

Sterling Silver. "Sterling" is one of the best known and the most respected markings in use. Pure (fine) silver, like pure gold, is very soft. Sterling silver is made of 92 1/2 per cent silver with 7 1/2 per cent copper added to give stiffness and wearing qualities.

Coin Silver. Coin silver, the alloy used in making United States silver coins, consists of 90 per cent silver and 10 per cent copper.

Silver Melting Points. Fine silver melts at 1762°F., sterling from 1650 to 1675°F., and coin silver from 1675 to 1690°F.

How to Order Silver. In ordering silver be sure to specify the width, as well as the gauge (thickness), desired. Silver can be purchased in almost any width and gauge. Also specify whether sterling or fine silver is wanted.

Fig. 115.—Hand rolling mill.

The Brown and Sharpe gauge is used in measuring thickness of silver. It is advisable, unless a rolling mill is available with which to reduce the thickness, to keep several gauges of silver on hand. The author has found from experience in dealing with students that sheet silver kept in 18-, 20-, and 24-gauge thicknesses will meet the requirements for most jewelry making, insofar as sterling silver is concerned.

Silver wire is sold by the ounce and costs more than sheet silver, owing to manufacturing costs. Many different sizes of wire are used in making jewelry, although the 12, 15, 20, and 30 gauges, in round form, are the sizes used most.

Fig. 116.—Reducing wire with a drawplate.

Use of Drawplate. By the use of a drawplate, it is a simple matter to make any desired gauge by drawing, or reducing, wire of larger size. Drawplates may be purchased for drawing round, half-round, triangular, square, and rectangular wire.

WEIGHTS OF DIFFERENT GAUGE STERLING SILVER IN SHEET AND WIRE FORM

Brown and Sharpe gauge	Thickness, inches	Sheet, ounces per square inch	Wire (round), ounces per lineal foot
5	0.1819	0.9978	1.7110
6	0.1620	0.8886	1.3568
7	0.1443	0.7913	1.0760
8	0.1285	0.7047	0.8534
9	0.1144	0.6276	0.6768
10	0.1019	0.5588	0.5366
11	0.0907	0.4976	0.4256
12	0.0808	0.4431	0.3375
13	0.0720	0.3947	0.2677
14	0.0641	0.3514	0.2122
15	0.0571	0.3129	0.1683
16	0.0508	0.2786	0.1335
17	0.0453	0.2482	0.1058
18	0.0403	0.2210	0.0840
19	0.0359	0.1968	0.0666
20	0.0320	0.1753	0.0528
21	0.0285	0.1561	0.0419
22	0.0253	0.1390	0.0332
23	0.0226	0.1238	0.0263
24	0.0201	0.1102	0.0209
25	0.0179	0.0982	0.0166
26	0.0159	0.0874	0.0131
27	0.0142	0.0778	0.0104
28	0.0127	0.0693	0.0083
29	0.0113	0.0617	0.0065
30	0.0100	0.0550	0.0052

Sterling silver is 1.24 times heavier than brass. Sterling silver is 1.18 times heavier than copper. Fine silver is 1.01 times heavier than sterling silver. Ten-karat yellow gold is 1.13 times heavier than sterling silver. Fourteen-karat yellow gold is 1.27 times heavier than sterling silver. Eighteen-karat yellow gold is 1.46 times heavier than sterling silver. A circle is 0.7854 times as heavy as a square of same diameter.

Fig. 117.

Point on one end the wire to be reduced in size by filing or by hammering. Put the drawplate in a vise, the smaller side of the hole nearer you, with copper or other soft metal protecting it from the vise jaws. Insert the pointed end in the hole nearest its size.

Grip the wire with the drawing tongs and pull through the plate. Annealing is necessary after a few draws to keep the wire from breaking. No silver is lost in the reducing process, as the wire simply becomes longer as its size is reduced.

Fine Silver for Bezels. Fine silver, used for making bezels, may be purchased by the ounce in various widths and gauges. Either 26 or 28 gauge is a good thickness for bezel making. The widths of 1/8 inch and 3/16 inch are the two most used in bezel making, although the 1/4-inch width is sometimes used in making rings from sheet silver, where a higher bezel is needed. If desired, fine silver may be purchased in sheet form and cut into various widths as needed.

Cost per Square Inch. It is advisable, if dealing with students, to find out how much your sheet silver costs per square inch and how much the wire costs per inch or foot. Although you will buy it by the ounce, the student will not call for it in that manner. He will instead request a certain size of sheet or so many inches or feet of a certain gauge wire. By knowing the cost per square inch or lineal foot, you can eliminate much weighing and save time.

SEMIPRECIOUS STONES

Semiprecious stones, both cabochon and faceted, suitable for use in handmade jewelry may be obtained from various dealers by those who do not care to cut and polish their own. Prices on stones vary, depending upon the quality, color, quantity bought, etc.

One of the best ways to buy stones is to arrange to have the dealer send an assortment of various stones "on consignment." Then you are permitted to look them over, select what you want, and return the remainder, paying only for what you have selected.

Fig. 118.—The metric scale, used in measuring many gem stones.

Often one is able to secure seconds, which are excellent for student work, especially for the beginner. These stones will appear as first-class stones in many instances, but they have slight defects, often color alone, imperceptible to the untrained eye.

Many of the dealers in gem-cutting material cut and polish a large number of stones, especially cabochons.

The karat, which is equal to 3 1/16 grains troy weight, is the unit of weight for weighing precious and some semiprecious stones. For convenience the karat is divided into 100 parts, each part of which is called a point. Thus a stone weighing 65 points would weigh 0.65 karat.

THE LAPIDARY

PREPARING THE STONES—SOME DIFFERENCES IN GEM-CUTTING.

THE name by which the cutter and polisher of stones is called is derived from *lapis*, a stone, but it is generally understood to mean one who includes in his work the art of engraving small stones and thus creating them "gems" for the use of the jeweller. Obviously the stones which in their rough state, although in crystalline forms usually coated over with an incrustation, must be prepared before they are ready for the jeweller or setter of precious stones. The incrustation has to be removed before the worth of the stone can be ascertained, for then when cleared of its outer coating, flaws, should any exist, are revealed. The uncut gem presents a very different appearance from the stone rendered lustrous by facets cut upon its surface, and the shining reflections cleverly contrived.

Many ancient craftsmen were, probably, workers in precious stones and in metal, and operated upon both substances. Others were gem-cutters rightly so-called for they confined their labours to engraving stones and investing them with that peculiar charm which is attached to anything on which are mystic characters or symbols which the common people rarely understand. Many of these ancient artists cut for their patrons stones which had been obtained from various sources, and engraved them according to their instructions, and then returned them to be afterwards set by other craftsmen.

PREPARING THE STONES.

In the preceding chapter the story of the precious stones has been told—their formation, discovery, and the way into which they have been pressed into the service of man—and used in worship in Pagan and in Christian times. In its natural state as taken from its matrix the stone may and has been used for artistic purposes as well as jewellery, but the jeweller prefers to make use of the precious stone after it has passed through the hands of the lapidary, who cuts, polishes or engraves it according to his fancy or that of his clients.

The term or craft-name, lapidary, is applied to all cutters of stones—precious, and of commoner kinds. The lapidary is one skilled in the cutting and polishing of stones. The tools of this artist are few in number, the principal being a lathe or wheel necessary for grinding or polishing. The engraver's tools are varied, but very fine and delicate. It is true much beautiful work has been done by hand with scarcely any mechanical aid; when we examine the fine work done by ancient craftsmen with few opportunities it is difficult to understand how they achieved such marvels.

Eastern artists, however, have always been famous for their wonderful handiwork with few tools and very primitive apparatus. Their skill in the present day has been seen in Eastern bazaars and in exhibitions in this country where Oriental natives have given demonstrations of their ability. Drills were known as far back as 725 B.C. The Etruscans used the drill and quite early the Greek artists understood its use as well as that of fine engraving tools. There was a fraternity of gem-cutters in 1373, in Nuremberg, where a trade guild was in existence and even then exercised a beneficial influence upon the craft. Apprentices were not allowed to trade or work on their own account for six years, during which time they were expected to become proficient, and to be able to undertake the ordinary work of gem-cutting.

In a previous chapter reference has been made to the different facets of the diamond; of the brilliant and of the rose diamond, which it may be mentioned was so named because it was supposed to resemble a rose when cut; it does not flash like the brilliant, and is therefore not now so much in favour.

SOME DIFFERENCES IN GEM-CUTTING.

It may be well to make clear some of the differences in the work of the lapidary, and in the methods adopted to ensure the best results. There are many quite ordinary stones in small slabs, cut and polished, which are used in jewellery. They have little or no reflective powers, but are in themselves beautiful, and for their rich colourings and tints are chosen for mounting in gold and silver. In Scotch jewellery the agates and the cairngorms are attractive and need no added beauty other than their setting. The veins of the green malachite are enough for the artist without any engraving.

Such stones as the amethyst and the topaz need no chiselling, other than simple facets to set off the flat surface of the top of the stone. Diamonds, rubies, and emeralds, and such precious stones, are carefully cut by the lapidary according to the approved style which gives the greatest brilliance to that particular stone.

Very different, however, is the work on which the engraver is engaged. The subject selected after having been scratched with a fine tool upon the surface (the polish having been removed) is cut either intaglio or as a cameo. Most of the ancient gems were cut intaglio, that is the design was cut into the stone, and shown up below the surface which was usually polished flat. The effect of intaglio gem-cutting is of course the opposite of cameo which is cut into the surface in order to throw up the pattern or design. Seals and signets are almost invariably cut intaglio, and when impressed upon soft wax or other substance reverse the effect of the intaglio and show impressed the clever cutting of the experienced artist. The early Greek intaglios cut into hard gems are indeed marvels of skill and patience, and show the rare ability to transfer in miniature quite small details of sculpture—the intaglios of Ancient Greece showing the human form and deities personified are among the priceless gems of the art of a far-off race of artists.

CPSIA information can be obtained
at www.ICGtesting.com
Printed in the USA
LVOW11s0923160418
573627LV00001B/33/P